Light in a Socio-Cultural Perspective

Light in a Socio-Cultural Perspective

Edited by

Ruth Lubashevsky and Ronit Milano

Cambridge
Scholars
Publishing

Light in a Socio-Cultural Perspective

Edited by Ruth Lubashevsky and Ronit Milano

This book first published 2017

Cambridge Scholars Publishing

Lady Stephenson Library, Newcastle upon Tyne, NE6 2PA, UK

British Library Cataloguing in Publication Data
A catalogue record for this book is available from the British Library

ISBN (10): 1-4438-7907-X
ISBN (13): 978-1-4438-7907-1

CONTENTS

Part 3: Creation and Embodiment

Part 4: Conceptual Functions of Light

LIST OF FIGURES

Fig. 6.3: Moritz Oppenheim (1800-1882), "The Eve of the Day of Atonement," lithograph, *Bilder aus dem altjüdischen Familienleben* (Frankfurt: Keller, 1886), unnumbered page.
Fig. 6.4: Geskel Saloman (1821-1902), *The Blessing of the Sabbath Candles*, 1898, oil on canvas, 120 X 140 cm.; courtesy of the Jewish Community of Stockholm.
Fig. 9.1: Gustave Doré, *Me miserable! Which way shall I fly, Infinite wrath, and infinite despair?* Engraving, Illustrated for Milton's *Paradise lost*, Image courtesy of the Poetry Collection of the University Libraries, University at Buffalo, The State University of New York.

INTRODUCTION

What is light? This question is so profound that it demands the compounded wisdom of different realms of thought to address the various facets of its meaning and implications. A physicist will discuss the physical properties of light, such as speed, wavelength, and the wave-particle duality, whereas a biologist will venerate its foundational role in the creation and perpetuation of life. Yet, the principal significance of light traverses the confines of the material world and is mirrored in the nonphysical one.

Since the dawn of civilization, the concept of light has been central to cultural processes and formulations. Light was regarded not merely as a physical phenomenon, but also as a moral concept, charged – in relation to darkness – with dialectical meaning –good/bad, enlightened/barbaric, and so on. We are aware of the significant role played by the concept of light through a range of societal and cultural phenomena, beginning with the naming of historical periods (i.e., the Dark Ages, the Age of Enlightenment) through religious perceptions evident in texts and works of art that figure God as the source of light and contemporary scientific discoveries that acknowledge the importance of light in the shaping of human mental states.

The United Nations declared 2015 the International Year of Light, with the goal of illuminating the achievements of light-science and of stressing its importance to humankind. Ben-Gurion University of the Negev in Israel participated in the UN program and organized a range of activities and events for various types of audiences. Parallel to a rich program dedicated to science per se, the president of the university, Prof. Rivka Carmi, together with UNESCO, supported the establishment of a lively theoretical discourse, which probed the perception and conceptual function of light in cultural history and the arts. In April 2015 the University's Department of the Arts organized a large exhibition of works by Israeli artists around the concept of light (curated by Prof. Haim Maor), alongside an international symposium designed to provide a comprehensive interdisciplinary perspective on the theme.

The nine chapters in this volume are based on papers delivered at that symposium. They provide various perspectives on the concept of light derived from different disciplinary methodologies within the Humanities, including cultural history, philosophy and intellectual history, literature, and visual culture, and probe the concept of light across history, from the Middle Ages through to the present. The collection is divided into four thematic – cross-disciplinary and cross-periodic – units. The first, *Between Image and Truth*, points to the role of light in the relationship between vision and knowledge. In her essay on stargazing in the twenty-first century, Romi Mikulinsky surveys the transition from the act of tracing light to the admiration of data, defining both as parallel practices of stargazing. This shift from the physical to the virtual, or from the object to the image, she argues, symbolizes a contemporary phenomenon by which representation is perceived as truth. This idea of light – whether sunlight or computer-screen light – as an agent of vision, and subsequently of truth, resonates in Stephen Gregory's discussion of Descartes's theories of light and vision. Investigating the bilateral relationship between light and truth, he defines the Cartesian understanding of light as a metaphor situated between viewing and knowing.

The second unit, *Exposing Reality*, offers three interpretations of light as a means of revealing reality. Niharika Dinkar analyzes a sculpture in St. Paul's Cathedral that presents the unveiling through lighting of an image associated with the mysterious Hindu tradition. Through her analysis, she draws a parallel between light and reason and situates the work and its narrative as part of the Enlightenment project of the West, which promoted the reasonable and realistic over the mysterious and spiritual. The theoretic correlation between light and reality is evident in the visual arts throughout history, but it gained particular resonance in the nineteenth century through the work of the Impressionists. In a continuation of the Realist Movement that celebrated modern reality, Impressionism focused on light as an agent of vision and of truth. Merav Fima, analyzing the motif of light in *Madame Bovary* by Gustave Flaubert, who was part of the Realist Movement, suggests that the novel anticipated the Impressionist aesthetics. By comparing Flaubert's literary representations to Claude Monet's art, she identifies the former as a pioneer of literary Impressionism. Taking the light-truth discourse to a practical level, Ruth Lubashevsky-Amar reflects on the role of infrared technology in the process of attribution in art history. In her essay, which focuses on Lucas Cranach the Elder, she suggests a model of attribution based on a combination of visual analysis and the power of light-based technology.

The section on *Creation and Embodiment* comprises two chapters that delve into the theological realm of Jewish tradition and Kabbalah. Susan Nashman-Fraiman studies the feminine tradition of candle lighting in Judaism, arguing that this practice resulted in the formation of a feminine understanding of creation, related to the making of light by women. Whereas Nashman-Fraiman interprets the light-creation trajectory in a gender-related context, Sandra Valabregue-Perry turns to medieval theosophical Kabbalah to show how images of light functioned as embodiments of such concepts as creation, emanation, and prophecy. In a further look at the potential functions of light in cultural and intellectual history, the forth and concluding unit, *Conceptual Functions of Light*, offers two scholarly attempts to redefine light. Yair Barak's essay deals with the metaphysics of the thirteenth-century theologian St. Bonaventure. Following Bonaventure's perception of light as a formal component of the structure of the universe, Barak discusses light as a measure, defining it as the quantifier of the spiritual distance from God. In the volume's concluding essay, Rina Jean Baroukh offers another perspective on light as a structural element in a discussion of the novel *The Wedding Gifts* (2014) by Israeli writer Shimon Adaf. Pointing to the idea of light as its leitmotif, she acknowledges its thematic function in the novel, but argues that at the same time the concept is presented as a structure of continuity and is comparable to the concept of time.

As a whole, this volume is the culmination of a project that was designed to provide a theoretical backdrop for the scientific developments illuminated in The Year of Light. It is our belief that these developments cannot be fully comprehended outside their cultural context and without considering their resonance in the epistemological sphere. The nine essays that follow reflect our desire to provide the contemporary reader with a diverse and interdisciplinary appreciation of the theoretical cultural and historic aspects of light.

SHARRON HASS, "THEFT," FROM *DAYLIGHT*

TRANSLATION TSIPI KELLER[1]

In the middle of the night, in one of the folds of night
pivoting for many years
on one thought
the hand draws twelve doors on the wall
for how can you leave the house
within a house wherein each step forward
is also a step downward

In the middle of the night, in one of the folds of night
visions that fade in the sun leave behind
dark matter, the sorrow of unlived lives.
Even the Leopard, prince of Salina,
will succumb to it. I have read.
I keep reading the same books again and again
(they are yellow-eyed and their joy is contained) –
in search of a throat for my slowest writing ever,
for what is forbidden to behold

(Somewhere nature prepares to watch me,
to transform me if I dare fill up
empty lines of poetry, since I did behold)

There's a woman in the woman who wants what she does not want
and she walks in the heat wearing a red wool coat
toward the impossible, to set on fire what the fire has refused.
All that is lacking the sweetness of a double and a shadow
yields to the grandest of lusts, the lust for beauty,
and to what is granted to god alone –
to bind defeat with ecstasy, to stamp the mark of disgrace
on those who see and are unseen, and to sing

---

[1] *Daylight* was published in 2011 by Bialik Institute (Jerusalem), and was awarded the Bialik Prize in 2012.

"The terrorist Wafa-al-Bass undresses
exposing her 'explosive underpants'
she attempts to detonate the bomb and fails
yesterday in a designated area near the Erez Roadblock"

If the messenger were to sing to the king
who had slept with his mother and killed his father
and, like a monster-magician, made blood entanglements
disappear, would the king still have the strength to say
"This terrible creature of darkness is me" and would not
his spirit, in face of a singing strophe of fate,
be disfigured in the humiliation of pain

There is night in the noon of day.
What is seen at a glance may become a fact
of the spirit. A madness lurking in pity.
I bent over the newspaper and saw your defeat,
like Maria's, elevated to the miracle of my life,
your failed death a marvel to our eyes – here, the double
has been revealed in a land of no shadows, traveling
within the borders of its dreams, with a wound as offering

PART 1:

BETWEEN IMAGE AND TRUTH

CHAPTER ONE

FROM THE OVERVIEW EFFECT TO THE DATA SUBLIME: STARGAZING IN THE TWENTY-FIRST CENTURY

ROMI MIKULINSKY

Humanity must rise above the Earth – to the top of the atmosphere and beyond – for only thus will he fully understand the world in which he lives.
—Plato[1]

Only an orbital perspection allows us to become new constructors of the globe. In the orbital sphere the closedness of the world opens towards the endless universe.
—Peter Weibel[2]

The photographic overview effect was caused by the broadcast of the first images of the Earth from outer space in the late 1960s. These, I suggest, transformed our worldview. This article proposes that this effect is now being mirrored in society's preoccupation with data and its visualization. In support of this claim, I present several examples from contemporary Western culture and arts that demonstrate the co-dependency that exists between the explosion of information technology and human imagination, focusing specifically on stargazing, astronomy, and celestial imaging.

The human desire for self-knowledge has a long relationship with stargazing that dates back to ancient Babylon and classical Greece. Since the beginning of civilization, man has lifted his gaze to the sky, attempting to find meanings and omens to explain and predict natural phenomena and elucidate the human condition. Not only has knowledge of our own planet often been dependent on human understanding of other planets, but our knowledge of ourselves and our environment is dependent on the subsequent development of visual aids for astronomical imaging and communication technologies.

This tendency, I argue, has intensified in the information age, in which data has become both the material and the tool with which we comprehend our world and our relationship to it. The more we rely on new technologies, the more they replace the old-fashioned direct human observation of celestial objects. For example, decoding information captured and retrieved by computers has replaced the traditional tracing of light. Radio telescopes intercept radiation – information – at wavelengths imperceptible to the human eye, and computers drive optical telescopes with their ever-larger mirrors and use “active optics” to “continuously compensate for subtle changes in the balance of the mirror and in the atmosphere above the instrument.”[3]

The twentieth century saw increasingly rapid advances in the scientific study of stars. Direct observation of the night sky is often mediated by instruments easily available to amateur astronomers (e.g., cameras, telescopes, and binoculars), and astronomical research is now the work of teams of scientists and engineers that rely increasingly on the analysis of digital, rather than optical data. Our knowledge of our own planet and of the universe is thereby almost always mediated and depends on technologies that emerged to serve the needs of governments and their military and intelligence establishments.

This article begins with the campaign initiated in the mid-1960s by the *Whole Earth Catalog* founder Stewart Brand.[4] I discuss the meanings and paradoxes inherent in the one-Earth symbolism, which is reflected in these images’ central role in the aesthetics of the environmental movement and counterculture. More recently, fascination with the sky took a new direction given the multitude of objects overcrowding the sky above us, be they satellites, drones, or other surveillance machinery. I then proceed by describing two exhibitions that attempt to theorize the shift from “planetary thinking” to orbitization, as Peter Weibel suggests.

Thanks to the objects that orbit around us and capable of managing and governing information flows, new possibilities for seeing are now available. These complex relations of seeing and being seen are unique conditions created at the end of the twentieth century that have become more complicated in the twenty-first. We have become accustomed to “seeing with machines” and regularly look at our world through our screens, navigating between online maps, livestreams of CCTV and surveillance cameras, and of course Google Earth and Street View. We can now leave Earth and venture into deep space using datasets made up of a million galaxies imaged by the Sloan Digital Sky Survey,[5] or see billions-of-years-old light from galaxies through the Hubble Space Telescope.[6] Our growing reliance on data and computerized processing of information is also apparent in Galaxy Zoo[7] a citizen science project for planet discovery and galaxy research. This project, which is part of the

Zooniverse[8] platform, invites volunteers to classify distant galaxies or discover planets around stars by detecting light curve changes and to take an active role in distant planet research.

I then attempt to explain the growing reliance on data in our data-driven society and discuss data-art and visualization. I focus on two projects that bring together the act of looking at data and stargazing: the first, July Freeman's *We Need Us* (2014), is a real-time sonification and visualization based on the metadata produced by the users of Zooniverse. The second artwork is *Black Shoals: Dark Matter* (2015) where financial data is presented like constellations of stars, looking and behaving like celestial objects. These artworks thereby inquire into stargazing today and introduce new meanings into it, illustrating the potency of space imagery and the extent to which our society is permeated with a fascination with data.

Until the late nineteen sixties, pictures of planets were scarce. In fact, the first photographs of the Earth became available only around 1968. Nowadays, space imagery is virtually ubiquitous. Apple regularly includes images of galaxies on personal computer screens and shots of the Earth as default iPhone wallpapers.[9] *Wired* publishes a daily photo from space.[10] With NASA's public outreach growing ever stronger on social media channels and the opening up of their archives and databases to the public,[11] we have access to both images from space and data and metadata about their production. Art critic Jonathan Jones hails space imagery as "the most important art" made today, and celebrates these images' ubiquity as proof of the extent to which science communication has transformed the way in which we gain knowledge about the universe and how we see it.[12] In this paper, I review the transition from the tracing of light to the admiration of data and information gathered about celestial bodies.

"Why Haven't We Seen a Photograph of the Whole Earth Yet?": *Whole Earth Catalog* and Planetary Thinking

It all starts with an image, a missing image in fact. The birth of what we now call cosmological consciousness is tightly connected to the birth of environmentalism, a geocentric perspective gained by being able to see Earth in full. In the 1960s, Stewart Brand initiated a protest demanding the release of Earth images. Both the Americans and the Russians had already deployed manned spacecraft that hovered above the Earth. Their cameras were directed – at least according to public announcements - to the depths of space rather than backwards to the direction of Earth. This campaign started with a vision. One night in 1966, 28 year-old Brand, a Stanford graduate and environmentalist, sat on a rooftop in San Francisco and looked at the night sky. High on LSD, he felt that what humanity needed most in order to acquire a sense of shared destiny was an image of Earth as seen from the outside.

> I imagined going farther and farther into orbit and soon realized that the sight of the entire planet, seen at once, would be quite dramatic and would make a point that Buckminster Fuller was always ranting about: that people act as if the earth is flat, when in reality it is spherical and extremely finite, and until we learn to treat it as a finite thing, we will never get civilization right . . . and I figured a photograph—a color photograph—would help make that happen.[13]

Inspired by his vision, Brand set himself up as a one-man protest movement, and started passing out buttons saying, "Why haven't we seen a photograph of the whole earth yet?" around the University of California campuses. He sent the buttons to NASA officials, congress members, scientists, and prominent thinkers such as Marshall McLuhan and Buckminster Fuller. Brand's campaign may or may not have influenced NASA, but a few months later the Apollo 8 astronauts, the first human beings to observe Earth from outside its planetary orbit, photographed their home planet from a lunar orbit. This image, released on December 24, 1968, came to be known as "Earthrise" (see Fig. 1.1).

Fig. 1.1: Earthrise Revisited. December 24, 1968. NASA Apollo 8 photo by Bill Anders, data visualization courtesy Ernie Wright, NASA Scientific Visualization Studio.

Photo-historian Robin Kelsey claims that no one anticipated the implications of the whole-earth photographs more accurately than Brand. The founder of the *Whole Earth Catalog* and the Long Now movement placed "Earthrise" on the cover of the fall 1969 edition of the catalogue. Since there were no available photos of the Earth in its entirety before "Earthrise," the catalogue's inaugural edition bore a satellite image of the Earth. For Brand, Kelsey continues, "these images were symbolic of a new economy that would respect ecological limits and yet reward entrepreneurial verve. The Whole Earth Catalog, subtitled 'Access to Tools,' linked the image of the earth to new distributive systems, and its global approach to information and product dissemination foreshadowed the powers of the internet."[14]

The *Whole Earth Catalog* is considered to have foreseen the digital age. It is also an important landmark in the birth of environmentalism and globalism. As John Markoff puts it, the catalogue was "the internet before the internet. It was the book of the future. It was a web in newsprint."[15] Not only prophetic, the catalogue was also hugely successful: it was published regularly between from 1968 to 1972, and its last edition sold more than 2 million copies.[16] In 1972, it won the first U.S. National Book Award in the category of contemporary affairs. For Brand and his fellow editors and writers, the planetary idea became a means to articulate a new universalism; once the image of whole Earth began to circulate, it illustrated that Earth was a fragile global village. The perspective gained by that photograph altered human consciousness and sense of place in the universe, and the catalogue as a directory of ideas and tools combined individual empowerment and education with whole systems concepts. According to Fred Turner "Brand's reader enjoys the power of a god to survey the whole earth below him. The front cover of many editions of the Whole Earth Catalog featured an image of the earth seen from space . . . In the Whole Earth Catalog, cold war technocracy itself had granted its opponents the power to see the world in which they lived as a single whole."[17]

The fact that whole-Earth or one-Earth symbolism became so potent is also tied to the fragility of Earth as conveyed by the photos that show it in isolation. It was the first time man could see Earth from afar, and the gaze back to Earth became a reminder of the planet's finitude and ecological vulnerability. Kelsey reminds us of the deep ironies inherent in the whole Earth images, viewed in the context of the race to the moon: "an image anointed as the sign of a universal humanity and ecological fragility had emerged from an intensely militarized international rivalry and a space program every bit as environmentally profligate as the society that produced it."[18]

There are many paradoxes inherent in the Whole Earth Catalog. A central one is, of course, the desire to go beyond cultural and political boundaries that was a fundamental concept for counterculture.[19] Brand's vision and quixotic campaign for the release of Earth images indicates yet another paradox: the conflation of the journey to inner space and outer space. This, I argue, is the meeting point of imagination and fantasy intermingle with the progress of technology, science is filled with dreaming and overcoming accepted boundaries.

Jon Palmesino and Ann-Sofi Rönnskog consider the Apollo 8 image as reaffirming, securing and fixing the planet, "albeit acting as a call for direct action, intervention and urgent reorganization."[20] They read "Earthrise" as an image shaped by the scientific technology of vision as well as by the surveying and measurement impetus and the accounting processes of Western empiricism.[21]

"We are as gods and might as well get good at it",[22] wrote Brand in the introduction to the first Whole Earth Catalog. Brand's cosmic view of Earth his sense that expanded consciousness was required for the next step in the evolution of humanity, and his call for personal freedom are all expressed in his ambition to equip his readers with knowledge and tools to transform their world. All these are encapsulated in the "Blue Marble" image.

The Overview Effect and the Effects of Orbitization

In 1972, four years after "Earthrise" was snapped by Apollo 8, the famous "Blue Marble" image was shot by the Apollo 17 astronauts (see Fig. 1.2). "Blue Marble" lent legitimacy to both the space program and twentieth-century concerns for the Earth's ecosystems and became emblematic of the "total systems" cybernetic view.[23] Both "Earthrise" and "Blue Marble" became iconic images and their effect on viewers was one of both awe and promise, proving the viability of leaving the planet to explore the unknown.[24]

Significantly, the visibility of the planet as a whole also made one's gaze visible. After all, in looking at something that was one's own Earth, one was also looking at oneself. As cultural geographer Denis Cosgrove emphasized, these images "have become the image of the globe, simultaneously 'true' representations and virtual spaces."[25] Moreover, Cosgrove argues that because they adopt the "one-world" approach, these images are charged, since they also contain the backwards-looking gaze of the astronauts that took them. Thus, they serve as a reminder of our common humanity and the artificiality of geopolitical boundaries.

Fig. 1.2: The Blue Marble from Apollo 17. December 17, 1972. Credit: Image courtesy NASA Johnson Space Center Gateway to Astronaut Photography of Earth

The astronauts' gaze and the great impression that seeing Earth from space left on them – and on those who have never risen beyond Earth's atmosphere – are at the heart of Frank White's book, *The Overview Effect*.[26] The book argues that the goal of reaching out into space is not merely to use extraterrestrial resources. Rather, humanity is laying the foundation for a potentially new civilization that will eventually push the evolution of human society forward. The astronauts' backward gaze is therefore at the same time also that of a forward-looking human race. Moreover, space exploration may serve an even a higher purpose by performing a vital function for the universe as a whole.

This sense of self-transcendence is articulated in our optimism regarding the human capacity to influence not only Earth, but the entire universe. The planetary idea that environmentalist groups hailed shifted in the nineteen eighties with the rise of globalization and communication networks and the resulting "flattening" of the Earth. The desire to overcome the boundaries of the atmosphere and what Peter Weibel calls the "law of the Earth" was the subject of "The Orbital Age," a 1986 symposium curated by Peter Weibel in Linz. He opens the catalogue text prepared for the symposium by heralding the beginning of a new era:

> We find ourselves at the end of a planetary thinking, where our globe was considered the "Nomos" of history. This has resulted in creating a world-spanning, unifying order of Earth and Space. This global order, though, tended towards a dangerous universalization, towards a totalitary One-World, the World State, taking away any possibility for emigration. With the orbital sphere we are now leaving the "Nomos of Earth", the Laws of our planet Earth, and humanity is offered the chance of doing away with the dangers of planetarization and universalization of a mono-world and a mono-culture.[27]

Whereas the Whole Earth Catalog sought to collapse boundaries between people and enable individual projects to alter life on a global scale, in the era of orbitization, the scale became larger. The inhabitants of Earth now aspire to unite Earth and Space and to celebrate the multitudes of opportunities this new horizon offers. Weibel describes a "starwards shifting of the borders" attested to by some 1500 satellites moving in orbit at the time of the exhibition.[28] These "1500 orbital ambassadors" also testify to the dislocation caused by the opening up, if not the abolition, of global boundaries. He considers globalization as flattening Earth cultures and warns of the creation of one unified society – a "mono-world and a mono-culture."

According to Weibel planetary thinking has been overtaken by orbital consciousness, and the orbital era opens new possibilities for the human race; we can now go beyond stargazing and looking at the light of the stars and think about new planets to live on. The colonization of space could even bring about the restoration of the Earth by making it possible to relocate to space. The symposium sought to explore the new possibilities opened up by overcoming gravitation and "launch[ing] out into the cosmic choreography of planetary movement."[29]

Some ten years after "The Orbital Age," "SpacePlace: Art in The Age of Orbitization" was launched at ZKM, co-curated by Weibel and Philip Pocock. In the time between the symposium and the exhibition, many more new satellites had been placed in orbit and not only the globe but also globalization itself had become a marketable commodity. Pocock and Weibel argue that selling often takes place through a cryptic orbital marketplace virtually invisible to many and overseen by supra-orbitant bodies, usually the military.[30] Globalization is no longer our most far-reaching context, they write, "no longer the border guard containing our world. The global village [is] now surrounded by a territory that is orbital in nature. Its nucleus no longer residing at the system's geometrical centre but rather a collective of city centres, financial hubs, parliament forums, even art centres to which orbital culture both gravitates and attempts escape."[31]

"SpacePlace" included both space and orbital artworks physically exhibited in Munich and Karlsruhe, as well as an online and mobile platform featuring a database of related projects. The platform envisioned by Pocock invited the audience to contribute items, entries, and artwork to the database. This participatory platform is based on Web 2.0 principles and semantic technology, and each node in the network of users/contributors is connected to another, creating a constellation-like structure. Following legendary video artist and satellite art pioneer Nam June Paik's 1988 statement regarding the need to subvert the military use of satellites and use them for pacifist purposes, ZKM calls for active public participation and aspires to raise public awareness of space policies dictated by military considerations.[32] Pocock openly writes that "appealing as it does to the thinking eye and the curiosity that space art has always sparked," the exhibition called attention to broader space concerns by focusing on cultural and peaceful uses of satellites and spacecrafts. Orbitization, according to the curators, means leaving the Earth's boundaries and expanding man's reach to the near earth orbit as well as to deeper space.[33] Looking at the sky therefore acquired new meanings that address the needs of citizens of the twenty-first century.

Survey and Surveillance: Data Art and Mediated Stargazing

Twenty years have passed since "SpacePlace." In the interim, we have seen the emergence of the information age and information society. Now that orbitization has taken command of the market, globalization is visible, "like a product on the shelf," and its effects are felt on both economic and cultural levels, writes Pocock.[34] Satellites and other geospatial systems collect and disseminate data, and also create images of Earth that are in fact artificial composites of massive quantities of data collected remotely by satellite-borne sensors .[35] According to South African architect Laura Kurgan, these composites are located at the "intersection between physical space and its representation, virtual space and its realization", raising fundamental questions about them.[36]

These new techniques that create images of Earth and other planets wrought from data and often provided in real-time articulate the new trends and obsessions that have by now become commonplace – what Viktor Mayer-Schöberger and Kenneth Cukier call "the datafication of everything."[37] Thanks to the internet, social networking, and smartphones, more data is being collected and stored than ever before. By analysing huge amounts of information, it is possible to discover patterns and relationships that were inaccessible and even invisible to us before. Nowadays, almost anything can be datafied – from pressure points across a retail floor, through sentiments expressed in speech or tweets, to sleep patterns or jogging heartbeats measured by smartphones. Tiny movement sensors can be integrated in commercial or public buildings to generate insights about employees, customers, visitors – and of course, suspects – and biometric data can be easily collected, stored, analysed and mobilized in service of predictive algorithms.

Moreover, unique 21st century ways of seeing have been developed with the availability of recent technological breakthroughs such as Google Earth and Street View. Nowadays anyone with an internet-connected computer or device can collapse time and space. Furthermore, the new survey and surveillance systems evaluate and calibrate – and thereby accelerate – the very flows they set out to measure, as described by Palmesino and Rönnskog.[38] They do so by creating more data on the data accumulated and adding complexity and detail to the info-scape.

The advent of imaging technologies has had a great impact not only on the way we see our place in the universe but also on the way we ascribe meaning to outer space and its impact on both the individual and collective imagination. Imagining and re-imagining space is now for the most part dependent and almost inseparable from digital technology. Similar to the possibility of exploring the Earth from the comfort of one's home, planet discovery and "space travel" have been made accessible thanks to the Internet. And since data is at the heart of contemporary digital culture, there are also records and statistics of reading and viewing habits, of access and public interest in space-related websites, video channels and crowdsourcing platforms.

For artist Julie Freeman, data about looking and analysing data about distant galaxies is the locus of inquiry: she considers this as the conjecture of art and science, of data and looking. One of her projects is *We Need Us*, launched in 2014 on the digital art platform TheSpace.org (see Fig. 1.3).[39] Freeman is interested in "life data" – the name she gives to the real-time data from living things. Freeman's work explores how we can use data to translate nature and generate new perspectives. For her, *We Need Us* is a thought experiment designed to reveal currently unknown "living" qualities of data and metadata. An article written with her research group refers to the material qualities of data and to the aspiration to use raw material to "expose more knowledge than ever before" in order to "gain insight beyond expectations of the past."[40]

We Need Us illustrates this ambition by giving sound and image to metadata retrieved from the Zooniverse website. Featuring a score of websites that crowdsource scientific problems that cannot be solved by computers alone, Zooniverse invites users to participate in various projects such as NASA's mission to spot stars in the early stages of forming planetary systems, or characterize surfaces on Mars. Freeman created algorithms that manipulate data on the activity within the website. Zooniverse enables anyone who logs into it to classify large scientific datasets – every action is recorded – so that whenever a Zooniverse user clicks or swipes online their actions are documented. The code Freeman created for *We Need Us* captures, collects and exposes the real-time activity of the platform's users. She thereby explores the real-time metadata of the galaxy research platform and transforms the work behind the live research into art. Each time a volunteer helps classify data, the shapes and sounds of the work change in accordance to the trail of metadata produced. This, she hopes, will remind visitors of the humanity in technology – the data needs us, just as much as we need them.

Fig. 1.3 Julie Freeman, Galaxy Zoo© CC BY-NC 4.0 Julie Freeman

The final example I wish to discuss is Lise Autogena and Joshua Portway's *Black Shoals* – an ongoing artwork conceived in 2000 that has gone through several renditions and is still exhibited worldwide, as part of the exhibition "Big Bang Data."[41] *Black Shoals: Dark Matter* (see Fig.

1.4) is a stock market planetarium: viewers are invited to gaze at a computerized night sky projected onto a domed ceiling above their heads. The sky is in fact a live representation of the world's stock markets, with each star representing a traded company. The stars in the planetarium slowly move across the sky, forming constellations, galaxies or even black holes in response to market fluctuations.

Fig. 1.4: *Black Shoals; Dark Matter* by Lise Autogena and Joshua Portway, Somerset House 2016. Photo credit: Joshua Portway

According to art historian Julian Stallabrass, the planetarium metaphor "encourages viewers to think about people's ancient relation to the stars as harbingers of fate, and the myriad vain methods of reading them to predict the future."[42] He refers to this spectacle of financial data visualization as a display of the "data sublime", in which the viewer is carried off imaginatively on waves of ungraspable data, reminiscent of how people used to be awed by "gigantic majestic or apocalyptic landscape paintings, or still are by safe(ish) views of mountains, storms and the sea." It is the pulsing of the data packets, he writes, that gives the requisite feeling of vertigo before a display of unconceivable complexity and speed.[43]

The data sublime, I argue, has replaced the sense of wonder felt while looking at Earth images. Whereas the "Overview Effect" involved an understanding of the fragility of our planet and at the same time inspired a promise of self-transcendence and overcoming of boundaries, the data sublime express the contemporary fascination with and fetishization of computers. The new way of seeing through machines and our growing dependence on screens breeds a new addiction to measurability, to resolution and to optimization. In a way, we have given up on seeing independently and trust the computer's eyes and measurements more than we trust our own. We know that computers can bring us that we cannot see – radiation, information, light curves and statistics and probabilities about them – but there is a price to our disavowal of unmediated, independent seeing.

Conclusion

The more we instrument the world through sensors, mass measurement and computation, the more we come to depend on data. Data becomes the material and textual infrastructure through which we read the universe that is gradually becoming an object for us to manipulate. With data streams and images composed of data streams, we have become accustomed to looking at the representation as if it were "truth." Conflating the physical and virtual spaces, we became accustomed not to think about the relationship between the representation and the represented, in fact, for many the latter has become epistemically irrelevant.

Data visualization and art projects such as *We Need Us* and *Black Shoals* remind us of our age-old fascination with the night sky. Even though we now rely on machines to mediate the sky for us, and have grown accustomed to looking at the stars mainly through screens, the hold stargazing has on our imagination is still powerful. The sense of awe is maintained and with it, mixed feelings about our place in the universe and our worshipping of the machines we have built to better understand the world. On the one hand, we are as gods, as Brand wrote back in 1968, and on the other, the enormity of the universe evokes an immanent sense of humility that is replicated by the data sublime. Looking at the sky in an attempt to trace stars, solve mysteries or predict the future still enchants us and as technology continues to, develop it will surely acquire new meanings. We know not what the next phase of visual technology will bring, and whether man will keep looking at the sky above Earth or raise his gaze to the sky stretched above other planets. Nevertheless, the spectacle that is the night sky, filled with promise and mystery continues to captivate human imagination.

Notes

[1] Quoted in "Declaration of Vision and Principles." The Overview Institute. Available at: http://www.overviewinstitute.org/about-us/declaration-of-vision-and-principles. Accessed on July 28, 2016.

[2] Peter Weibel, "Introduction," *The Orbital Age*. Ars Electronica, 1986. Available at: http://90.146.8.18/en/archives/festival_archive/festival_catalogs/festival_artikel.asp?iProjectID=9204 . Accessed on July 28, 2016.

[3] Michael A. Hoskin, *The History of Astronomy: A Very Short Introduction* (Oxford: Oxford University Press, 2003), 110.

[4] Stewart Brand. "'Whole Earth' Origin..." sb.longnow.org/SB_homepage/WholeEarth_buton.html accessed August 10, 2016.

[5] Sloan Digital Sky Surveys, www.sdss.org/surveys/. Accessed August 10, 2016.

[6] Hubble Space Telescope, http://hubblesite.org/gallery/. Accessed August 10, 2016.

[7] Galaxy Zoo, https://www.galaxyzoo.org. Accessed August 10, 2016.

[8] Zooniverse, https://www.zooniverse.org. Accessed August 10, 2016.

[9] See Jeff Richardson, "Blue Marble," *iPhone J.D.*, March 10, 2010. Available at: http://www.iphonejd.com/iphone_jd/2010/03/blue-marble.html. Accessed August 10, 2016.

[10] See "Wired Space Photo of the Day," *Wired*, www.wired.com/tag/space-photo-of-the-day/. Accessed August 10, 2016.

[11] See "Social Media at NASA," National Aeronautics and Space Administration. www.nasa-usa.de/socialmedia. Accessed August 10, 2016.
Mark Boyd, "NASA Launches API Portal to Encourage Use of Data Sets," *ProgrammableWeb*. May. 26, 2015. Available at: www.programmableweb.com/news/nasa-launches-api-portal-to-encourage-use-data-sets/2015/05/26. Accessed August 15, 2016.

[12] Jonathan Jones, "Out of this World: Why the Most Important Art Today is Made in Space," *The Guardian.* June 12, 2016. Available at: https://www.theguardian.com/artanddesign/2016/jun/12/why-most-important-art-today-made-in-space. Accessed August 25, 2016.

[13] Stewart Brand, "Photography Changes our Relationship to Our Planet," *Click! Photography Changes Everything*, a Smithsonian Photography Initiative:

https://web.archive.org/web/20090401220327http://click.si.edu/Story.aspx?story=31. Accessed July 20, 2016

[14] Robin Kelsey, "Reverse Shot: Earthrise and Blue Marble in the American Imagination," in *New Geographies, 4 Scales of the Earth*, ed. El Hadi Jazairy (Cambridge: Harvard GSD, 2010), 13.

[15] Carole Cadwalladr, "Stewart Brand's Whole Earth Catalog, the book that changed the world," *The Observer*, May 5, 2013. Available at: https://www.theguardian.com/books/2013/may/05/stewart-brand-whole-earth-catalog. Accessed July 21, 2016.

[16] The Museum of Modern Art, "Access to Tools: Publications from the Whole Earth Catalog, 1968-1974," Interactive exhibition. Available at: http://www.moma.org/interactives/exhibitions/2011/AccesstoTools/. Accessed September 5, 2016.

[17] Fred Turner, *From Counterculture to Cyberculture: Stewart Brand, the Whole Earth Network, and the. Rise of Digital Utopianism* (Chicago: University of Chicago Press, 2006), 83.

[18] Kelsey, 12-13.

[19] Diedrich Diederichsen and Anselm Franke (Eds.), *The Whole Earth California and the Disappearance of the Outside* (Berlin: Sternberg Press, 2013).

[20] John Palmesino and Ann-Sofi Rönnskog, "Plan the Planet", in *The Whole Earth: California and the Disappearance of the Outside*", eds. Diedrich Diederichsen and Anselm Franke (Berlin: Sternberg Press, 2013), 83.

[21] Palmesino and Rönnskog, 82.

[22] Stewart Brand, "We are As Gods," Whole Earth Catalog, Fall 1968. Available at: http://www.wholeearth.com/issue/1010/article/195/we.are.as.gods. Accessed July 22, 2016.

[23] Volker M. Welter, "From Disc to Sphere Taking on the Whole Earth", *Cabinet*, no. 40 Hair (Winter 2010/11). Available at: www.cabinetmagazine.org/issues/40/welter.php. Accessed July 22, 2016.

[24] According to Welter (ibid.), the earliest photographs of the Earth were far from being a spectacle and were not all that impressive. They showed semi- or near-circular shapes set against streaky black-and-white backgrounds. He notes that until photographs of the entire earth were technically achievable, the limited field of view available from lower altitudes meant that the frame of the photograph dissected the planet into random pieces.

[25] Quoted in Laura Kurgan, *Close Up at A Distance: Mapping, Technology, and Politics* (New York: Zone Books, 2013), 12.

[26] Frank White. *The Overview Effect: Space Exploration and Human Evolution* (Boston: Houghton Mifflin Harcourt, 1987).

[27] Peter Weibel, "Introduction," *The Orbital Age*. Ars Electronica, 1986. Available at: http://90.146.8.18/en/archives/festival_archive/festival_catalogs/festival_artikel.asp?iProjectID=9204 . Accessed July 28, 2016.

[28] According to reports made by the United Nations Office for Outer Space Affairs (UNOOSA), at the end of August 2015, there were 4077 satellites orbiting the Earth. Of the 4,077 satellites in orbit only 1,305 are active report The Union of Concerned Scientists (UCS). Andrew Lavender, "How many satellites are orbiting

the Earth in 2015?" *Pixalytics*. 23 September 2015. www.pixalytics.com/sat-orbit-2015/. Accessed July 30, 2016.

[29] Peter Weibel, "Beyond the Earth: The Orbital Age," Ars Electronica 1986. Available at: 90.146.8.18/en/archives/festival_archive/festival_catalogs/festival_artikel.asp?iProjectID=9206. Accessed July 20, 2016.

[30] Phillip Pocock and Peter Weibel, *SpacePlace*: *Art in the Age of Orbitization* (Karlsruhe: ZKM Center for Art and Media, 2006), 51.

[31] ibid., 52.

[32] ibid., 50.

[33] Philip Pocock, "Look Up! Art in the Age of Orbitization." *Imagining Outer Space: European Astroculture in the Twentieth Century,* ed. Alexander C. T. Geppert (Basingstoke, UK: Palgrave Macmillan, 2012), 328.

[34] Pocock, 328.

[35] The term "remote sensing" is used to describe the science and art of identifying, observing, and measuring an object without coming into direct contact with it. According to NASA's website, "this process involves the detection and measurement of radiation of different wavelengths reflected or emitted from distant objects or materials, by which they may be identified and categorized by class/type, substance, and spatial distribution." Steve Graham, "Remote Sensing", Earth Observatory, 17 September, 1999. Available at: http://earthobservatory.nasa.gov/Features/RemoteSensing/remote.php. Accessed August 20, 2016.

[36] Kurgan, 12

[37] Viktor Mayer-Schöberger and Kenneth Cukier, *Big Data: A Revolution That Will Transform How We Live, Work, and Think* (Boston: Houghton Mifflin Harcourt, 2013).

[38] Palmesino and Rönnskog, 82.

[39] Julie Freeman, *We Need Us*, *The Space.org*. Available at: https://www.thespace.org/artwork/we-need-us. Accessed on August 20, 2016.

[40] Julie Freeman, Geraint Wiggins, Gavin Starks, and Mark Sandler. A Concise Taxonomy for Describing Data as an Art Material. Proceedings of IEEE Vis, Chicago, 2015. Available at: http://visap.uic.edu/2015/VISAP15-Papers/visap2015_Freeman_ConciseTaxonomy.pdf. Accessed August 20, 2016.

[41] "Big Bang Data", curators Olga Subirós and José Luis de Vicente. Somerset House, Available at: http://bigbangdata.somersethouse.org.uk. Accessed August 20, 2016.

[42] Julian Stallabrass, “A View from the Fish Tank,” *Black Shoals* website. Available at: http://www.blackshoals.net/writing/. Accessed August 20, 2016.

[43] Stallabrass, ibid.

CHAPTER TWO

CARTESIAN ILLUMINATIONS: TRANSFERS OF LIGHT BETWEEN THE PHYSICS AND PHILOSOPHY OF DESCARTES

STEPHAN GREGORY

Illumination, Divine and Profane

The philosophical writings of Descartes are characterized by a vivid metaphorical exchange between the concepts of truth and certainty on the one hand, and the field of visual perception on the other. A metaphorological analysis as demonstrated by Hans Blumenberg in his seminal essay "Light as a Metaphor for Truth"[1] may show how notions originating from the field of sensual light experience migrate to theological and philosophical conceptions of truth, thus constituting a kind of "aesthetic subconscious" of the rational order of thought. Examining the processes of metaphorical intercommunication between the realm of everyday experience and philosophical reasoning can be especially instructive in the case of Descartes, since he is known for his attempt to free philosophical thinking from all that is metaphorical or ambiguous.

Being the most prominent of Cartesian light metaphors, the "natural light of reason" may be worth a closer look. Within the Cartesian epistemology, this *lumen naturale* acts as a sort of indicator light, a luminous index of truth, spreading out instantaneously once a true cognition has been attained. "A great light in the intellect," Descartes notes in his *Meditations*, is "followed by a great inclination in the will"[2] to accept a judgement as true. The image of light is associated here with the idea of the immediacy, involuntariness, and irresistibility of truth. If we take such a metaphorical use of light as an act of translation from one sphere to another, this operation includes not only the transmission of an image, but also a transfer of power. Along with the apparition of light, the persuasive potentials inherent in the religious truth of revelation are transferred to the field of subjective cognition and the power of illumination, which has always been above all reason, is now attributed to reason itself.

This kind of transaction naturally bears the danger of blasphemy. St. Augustine still counters the idea of man considering himself as light with a triple ban: “lumen tibi esse non potes; non potes, non potes.”[3] Therefore, in scholasticism the term *lumen naturale* is only used with caution. And still Descartes tries to reconcile the natural light of reason with theological doctrine by characterizing it as a God-given *facultas*. On the other hand, he grants it considerable autonomy. From Descartes on, it is no longer the light of divine illumination that guides the human practices of knowledge. What counts instead now is “the indubitable conception of a clear and attentive mind which proceeds solely from the light of reason.”[4]

Thus, the self-empowerment of reason can well be referred to as a change of light. It is described as such in Foucault’s *Folie et Déraison: Histoire de la folie à l'âge classique*: The rigorous division between reason and madness is accompanied by another division concerning the ways of being-in-the-light: “[Man] was to learn to dominate that great division, and bring it down to his own level; and make in himself the day and the night, and order the sun of the truth to the pale light of *his* truth.”[5] Foucault mentions not only a difference in light intensities (the dazzling, glistening light of Divine Revelation versus the “weak” light of the human mind), but also different types of light. The *lumen naturale* turns out to be precisely a non-natural, artificial light – one that allows man to switch arbitrarily between day and night. This conception of light as a regulated and, thus, artificial brightness corresponds to the widely-held view that pure light would leave things just as invisible as pure darkness. According to the optical *episteme* of the time, light can only serve as a medium of visibility if it is dimmed and mingled with shadows: “Just as darkness [*le tenebre*] stays invisible if it is not partially alleviated by light, pure light . . . is not visible to the human eye either.”[6]

Thus, the revolution in the order of truth, which began in the seventeenth century and has managed to place knowledge (in the shape of subjective certainty) above all other forms of cognition, may be referred to as a revolution of the order of light. To the extent that truth is no longer understood as something that appears by itself, but must be revealed and wrested from its concealment,[7] the character of light changes, too. The depreciation of the occurring, outer light (with all its dangers of deception) comes along with the empowerment of a man-made, controllable and (re)presentational light. The artificial and technical character of this new light can hardly be hidden by calling it *lumen naturale*.[8]

Two-way Traffic

Beyond Blumenberg's approach, the metaphorical traffic between light and truth can also be analysed in the opposite direction. Instead of only asking how the sensual experience of light contaminates the conceptual definitions of truth and knowledge, we might also try to find out what happens to light when it is used as a metaphor and, thus, as a means of transport for truth. So the question is not only "How does truth adjust to light?" but also: "How does light adjust to truth?" To what extent are not only theories, but also the material usages and instruments of light affected by their involvement in different praxeologies of truth, dramaturgies of illumination, or scenarios of cognition?

As I would like to show, it is not light as such that serves as a metaphor for knowledge in the Early Modern Age. In fact, we are dealing with a very specific, technically configured form of light. In order to work as a metaphor for the new kind of truth, light first has to be configured and rectified in a particular way. This means, on the one hand, that those qualities of light that smoothly fit into the new order of certainty (for example, the linear spreading and the geometric calculability of light) have to be underlined and reinforced. On the other hand, there are some aspects of light, such as colour, gloss, diffraction, diffusion, and so on, that must be suppressed or discarded because they interfere with its usability as a metaphor for knowledge. In the following sections, this hypothesis will be tested on the basis of Descartes's physical theories of vision and light. To what extent can we find here a tendency to adjust vision and light to the requirements of Cartesian epistemology, to "format" them in such a way that they comply with the ideals of truth and certainty?

Cartesian Vision

Descartes's *Dioptrique* (1637, English titles: *Optics* or *Dioptrics*) can be considered a mechanistic theory of vision. It represents the attempt to explain optical imaging and visual perception exclusively by the way in which matter interacts with motion.[9] With respect to anything happening before the retina, Descartes relies on the *camera obscura* model of vision known from Kepler's book *Ad Vitellionem Paralipomena* (1604).

As far as this external part of visual perception (illustrated as a geometric point-to-point-projection) is concerned, a "similitude" of representation can indeed be observed. Descartes emphasizes how perfectly the outside world is "depicted" on the back of the eye: "But you will be even more certain of this . . . taking the eye of a newly deceased man, or, for want of that, of an ox or some other large animal . . . . Now . . . it will be easy for you to understand that if you are enclosed in the chamber P, and fix your eyes on the white body RST, you ought to see there the likeness of the objects V, X, Y."[10] Descartes's optical theory of vision ends with this description of the retinal image. All the rest of the process of vision, the further processing of images between retina and *res cogitans*, is subject to laws that have nothing to do with the idea of mimetic reproduction.

As Descartes emphasizes, mental images do not necessarily need to resemble the outside world in order for the mind to accept them as true representations. Quite the contrary is true; a representation is often more truthful if it is not completely similar to its object. Descartes illustrates this point with the example of copper plate engraving, a representational technique that seems especially suited to providing a useful, simple picture of reality precisely because of its low mimetic fidelity, i.e. due to its omissions and reductions.[11]

Cartesian Light

Descartes's *Dioptrique* (1637) (fig. 2.1) is not addressed to philosophers, but rather to craftsmen, lens-grinders and manufacturers of optical instruments. Its explicit aim is to help human perception overcome its natural limitations by means of technical enhancement.[12] In accordance with this pragmatic approach, Descartes does not even bother to question the nature of light. Instead he goes directly to the investigation of concrete optical phenomena, such as light refraction.[13] However, the explanation of these phenomena turns out to require at least the introduction of certain model-like assumptions about light.

The best-known of these models is probably the tennis analogy, in the context of which the reflection and the refraction of light are described according to the mechanical models of recoil effects and, respectively, surface penetration. Whereas here the comparison with tennis balls suggests a particle theory of light, Descartes elsewhere argues – this time by drawing an analogy with a blind man's stick (fig. 2.2)– that it is not the particles that constitute light, but the motional impulse that runs through them.

The mechanical impact that occurs at the outer end of the cane is transmitted to the hand of the blind person without any temporal delay. In an analogous manner, it is suggested that the propagation of light should be pictured as an impulse of movement being passed on by mechanically coupled particles.

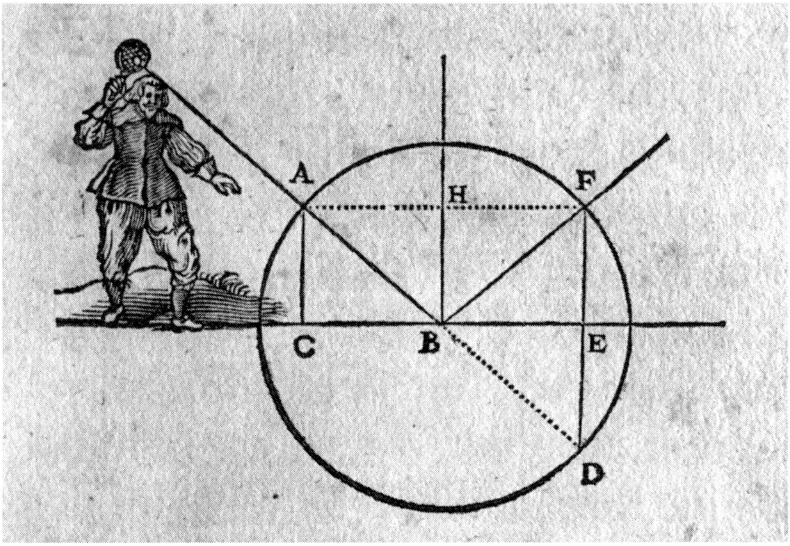

Fig. 2.1: The Tennis Ball Model, from René Descartes: *Discours de la méthode* [...] *plus la dioptrique,* Leyden, 1637.

Fig. 2.2: The Cane Analogy, from René Descartes: *Discours de la méthode* [...] *plus la dioptrique,* Leyden, 1637.

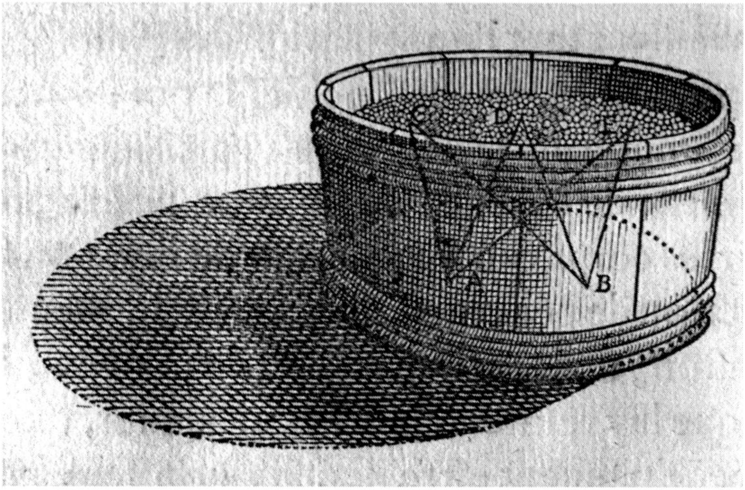

Fig. 2.3: A vat filled with grapes, from: René Descartes, *Discours de la méthode* [...] *plus la dioptrique,* Leyden, 1637.

A further illustration, a "vat at vintage time, completely full of half-pressed grapes," (fig. 2.3)[14] also shows the deficiencies of the model. If the holes A and B in the bottom of the vat (which stand for the eyes of an observer) are opened, the grape and fluid particles will doubtlessly rush towards the openings. Yet, the resulting chaos of particles will hardly be suited to imparting the idea of a linear, instantaneous transmission of a motional impulse.

The Idea of True Vision

The fact that Descartes's conceptions of light and vision are modelled according to the epistemological ideal of certainty becomes especially obvious in the way Descartes distinguishes between optical qualities that imply a "clear and distinct" (and therefore truthful) perception and those qualities that do not imply such perceptions and are thus incapable of indicating truth.

As for visual cognition, this difference between optical qualities that do or do not imply truth becomes manifest in the opposition of shape and colour. The reason why the perception of shape is a superior form of cognition is that it follows "a natural geometry."[15] Since, in this case, vision is considered a kind of scanning method that uses light beams as rulers, it becomes comparable with the triangulation methods of land surveyors. Accordingly, Descartes stresses the basic substitutability of vision by the tactile sense; everything we can focus our eyes on could just as well be sensed with correspondingly extended sticks – which is also what guarantees the exactness of that form of cognition.[16]

Thus, shape recognition is not only superior to colour perception (which cannot be grasped by simple geometric deduction). It also has the advantage that it can be technically enhanced. Size perception, for instance, is imperfect, but its deficiencies can be compensated for artificially. All this results in a programme of technical vision designed to supplement or substitute for the deficient performance of the human senses with the reliable services of artificial organs.

The Truth of Light

It now becomes evident that all the aspects of vision that are capable of truth, i.e. those that are perfectly determinable in geometric terms, are actually not of a specifically visual character. They could be replaced, for instance, by an enhanced tactile sense. So one could argue that the Cartesian epistemological programme pursues a "de-visualization of vision," since Descartes seems to reduce the process of visual perception to a mechanism of localization that is indifferent to whether its means are eyes, sticks, or tensioned threads.[17] For the same reason, it would be legitimate to characterize Descartes's notion of visual certainty in terms of a "seeing without light", i.e., as a kind of vision that is precisely not determined by the medial specificities of light.

What could, after all, a "Cartesian image" look like? What are the concrete attributes of a visual representation that complies with the postulates of clarity and distinctness, and, therefore, is suited to show the truth? First, this representation would be colourless, for colour is only confusing, discloses nothing about objective reality, and distracts from the question of material extension, which is, after all, the only question that permits reliable judgements.[18] Second, the required form of depiction would grant higher priority to the exact geometric localization of an object than to capturing an impression of light. Thus, graphic techniques are to be preferred over paintings, given that the latter work with surfaces and shadings. Here it is instructive to take a second look at Descartes' example of engravings. In contrast to, for instance, etching techniques, engravings only know black and white – only lines, no surfaces or dots. This is what determines that they will, more than any painting, capture definite contours and give quasi-geometric locational descriptions. In view of these considerations, the preliminary conclusion could be drawn here that in spite of all the pictorial metaphors, Descartes represents an utterly unpainterly concept of vision.[19] Thus, following Wölfflin's famous distinction between "linear" and "painterly" images,[20] the Cartesian pictorial programme would have to be located on the side of the "linear."

Does this therefore mean that Descartes is the one who forever hinders the emergence of light by submitting it under the primacy of representation and forcing it into a pre-measured and pre-construed space of cognition?[21] Even though this notion would fit well into the widespread image according to which Descartes has to be made responsible for all the misadventures of modern subjectivity, we can plead for a pardon in this special case. On the one hand, Descartes's description of the visual process may indeed be called reductive, as it blinds out the characteristic mediality of light for the sake of geometric simplicity. Nevertheless, Descartes has also sketched a physics of light, which, inspired by models of hydromechanics, gains remarkable momentum of its own.

In the framework of the hydromechanical model, Descartes understands light as a motional impulse, induced by a rotating source of radiation and paving its way through a universe crammed with particles. The image of a vat filled with grapes, as it appears in Descartes' *Optics*, makes it difficult to imagine a linear propagation of this motional impulse.

Yet, it becomes downright impossible when we consider how Descartes imagines air particles mingling with ether particles, or when we look at the illustration of cosmic vortexes in his *Principia Philosophiae* (fig. 2.4). Considering this chaotic swarm of particles, how is the *actio*, i.e., the motional impulse of light, supposed to be passed on in a straight line?

As Descartes himself remarked, there is a noticeable divergence between the ideal geometric nature of light propagation and its medial reality. For instance, regarding the lines that indicate the direction of the motional impulse (which for Descartes are "in fact light rays"), "we should not stop thinking of them as perfectly straight."[22] At the same time, however, Descartes presumes that "the parts of the second element that serve to transmit this action . . . can almost never . . . make exactly straight lines."[23] Here, we might say that physics puts a spoke into the wheel of geometry. Descartes's theory of light propagation is meant to make light effects entirely predictable in mathematical terms, but the multiplicity of motions that come into play makes such a prediction seem practicably impossible.

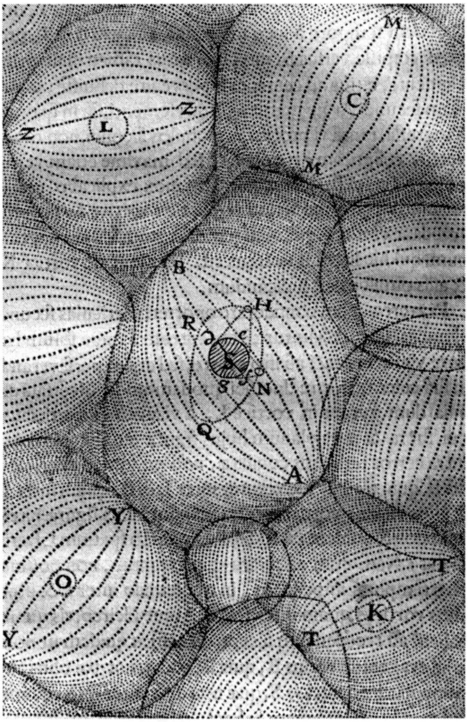

Fig 2.4: A Vortex in heavenly matter, from René Descartes, *Principia Philosophiae*, Amsterdam 1644.

Regarding his epistemological programme, Descartes is indeed an advocate of the "linear," and can therefore be termed a "man of the Renaissance," as Gilles Deleuze has done.[24] His physics of light, however, is best understood as the work of a thinker who, although involuntarily, shows a certain affinity to the "painterly," i.e., to the Baroque. The luminous subversion of the geometric line in Descartes' physics presents the opportunity to re-think the relationship between the Cartesian project and the pictorial programme of the Dutch painters.

Of course, a clear convergence can be observed between the "art of describing,"[25] which emerges in seventeenth-century Holland, and Descartes's claim for a clear and distinct representation of external reality. It also seems plausible to establish a link between the experiments of Dutch landscape painters and the Cartesian theory of vision, especially with respect to Cartesian theory's proposals regarding the technical enhancement of the human eye.[26] Seventeenth-century Dutch painting and Descartes share the idea of artificially regulating vision for the sake of true representation.

There is, however, another point, which cannot quite as easily be reconciled with the programme of geometrical domestication. The Cartesian physics of light and the corresponding presumption of a swarm of particles being traversed by rotational and directional impulses is hardly compatible with the ideal of an exact geometric representation of the outside world. Yet, it is closely related to the practical intuitions of those artists who, living next-door to Descartes during his self-chosen Dutch exile, invented a new approach to light in painting.

Rembrandt's drawings and paintings in particular show a remarkable proximity to Descartes' conception of light. This is not only the case for the choice of subjects in some of Rembrandt's works;[27] it also becomes manifest in his treatment of light.

Instead of conceiving of light merely as "lighting,", that is, as a source of visibility, Rembrandt appreciates the physical presence of light, making its mediality obtrude as a visual appearance of its own. As historian of science Alan E. Shapiro puts it, in Rembrandt's paintings "light seems to be a palpable presence, it exudes colour and does not merely reveal it, and surfaces glow, as if they were pushing outwards."[28]

In the present context, I am not so concerned with the question of whether it was philosophy that influenced painting, or painting that inspired the philosophical conceptions of light, in any case a question that could hardly be resolved. Perhaps it is enough to notice that in seventeenth-century Holland two very similar interactions between light and truth simultaneously took place on two different levels. It seems as if light itself had shown some kind of resistance against its epistemological domestication. While rational philosophers believed that light was an obedient instrument of representation, something else, something that lies beyond instrumental exploitability appeared on the horizon. Instead of being reduced to a medium of truth, light shows itself in its own truth, in the truth of its mediality.

Notes

[1] See Hans Blumenberg, "Licht als Metapher der Wahrheit," in *Ästhetische und metaphorologische Schriften*, ed. Anselm Haverkamp (Frankfurt am Main: Suhrkamp, 2001). The citations in this essay are taken from the English translation: Hans Blumenberg, "Light as a Metaphor for Truth: At the Preliminary Stage of

Philosophical Concept Formation," in *Modernity and the Hegemony of Vision*, ed. David M Levin (Berkeley: Univ. of California Press, 1993).

[2] René Descartes, *Meditations on First Philosophy: With Selections from the Objections and Replies*, ed. John Cottingham (Cambridge: Cambridge Univ. Press, 1996), 41.

[3] Augustinus, Sermo CLXXXII, quoted in Hans Blumenberg, "Light as a Metaphor for Truth, 43.

[4] René Descartes, "Rules for the Direction of the Mind," in *The Philosophical Writings: Translated by John Cottingham, Robert Stoothoff, and Dugald Murdoch* vol. 1 (Cambridge: Cambridge University Press, 1984), 14.

[5] Foucault, Michel. "Preface to the 1961 edition [of *Folie et déraison*]," in *History of madness*, ed. Jean Khalfa, (London: Routledge, 2006), xxxiv.

[6] Carlo Antonio Manzini, *L'occhiale all'occhio: Dioptrica pratica* (Bologna: per l'herede del Benacci, 1660), 41f. (my translation).

[7] See. Hans Blumenberg, "Light as a Metaphor for Truth," 52.

[8] Blumenberg indicated that, with the transition from God-given to man-made light, increasingly "technological figures come to invade the metaphorics of light." (ibid., 170)

[9] See Nancy L. Maull, "Cartesian Optics and the Geometrization of Nature," *The Review of Metaphysics* 32, no. 2 (1978): 253.

[10] René Descartes, "Optics," in *Discourse on Method, Optics, Geometry, and Meteorology.*, trans. Paul J. Olscamp, Rev. ed. (Indianapolis, IN: Hackett, 2001), 91, 94.

[11] See ibid., 90.

[12] See Neil M. Ribe, "Cartesian Optics and the Mastery of Nature," *Isis* 88, no. 1 (1997): 44.

[13] See Claus Zittel, *Theatrum philosophicum: Erfahrungsmodi und Formen der Wissensrepräsentation bei Descartes* (Berlin: Akademie-Verlag, 2009), 280.

[14] Descartes, "Optics," 69.

[15] Ibid., 106.

[16] See René Descartes, "Discourse on The Method for Rightly Directing One's Reason and Searching for Truth in the Sciences," in *Discourse on Method, Optics, Geometry, and Meteorology*, 67: "True, this sort of sensation is rather confused and obscure in those who do not have much practice with it, but consider it in those who, being born blind, have made use of it all their lives, and you will find it so perfect and so exact that one might almost say that they see with their hands . . . ."

[17] See Dalia Judovitz, "Vision, Representation, and Technology in Descartes," in Levin, *Modernity and the Hegemony of Vision*, 68.

[18] In fact, this is the aesthetic lesson French Classicism would learn from Descartes: priority of form over colour, as colour is considered a menace to truth. See Stephen Gaukroger, *Descartes: An Intellectual Biography* (Oxford, New York: Clarendon Press; Oxford University Press, 1995), 4.

[19] See Judovitz, "Vision, Representation, and Technology in Descartes," 84.

[20] See Heinrich Wölfflin, *Kunstgeschichtliche Grundbegriffe: Das Problem der Stilentwicklung in der neueren Kunst*, 5. Aufl. (München: Bruckmann, 1921).
[21] See Martin Heidegger, "Die Zeit des Weltbildes," in *Holzwege* (Frankfurt am Main: Klostermann, 1980), 85.
[22] René Descartes, "Treatise on Light and Other Principal Objects of the Senses," in *The World and Other Writings*, ed. Stephen Gaukroger (Cambridge: Cambridge Univ. Press, 1998), 64.
[23] Ibid.
[24] Gilles Deleuze, *The Fold: Leibniz and the Baroque* (Minneapolis: University of Minnesota Press, 2007), 32.
[25] See Svetlana Alpers, *The Art of Describing: Dutch Art in the Seventeenth Century* (Chicago: Univ. of Chicago Press., 1983).
[26] See Erik Larsen, "Descartes and the Rise of Naturalistic Landscape Painting in 17th Century Holland," *Art Journal*,24, no. 1, Autumn (1964): 12–17.
[27] See Cf. British Museum, „*Rembrandt – The Three Trees*",, print, (The British Museum, London) cCurator's comments, retrieved from http://www.britishmuseum.org/research/collection_online/collection_object_details.aspx?objectId=756785&partId=1 (accessed February 20, 2017). Here the assumption is made that the depiction of light, clouds, and rain in Rembrandt's *Three Trees* is inspired by an illustration in Descartes *Les Méteores* (1637).
[28] Alan E. Shapiro, "In Search of Cartesian Painting: Descartes on the Nature of Light and Color," in *Lichtgefüge des 17. Jahrhunderts: Rembrandt und Vermeer - Spinoza und Leibniz*, ed. Carolin Bohlmann, Thomas Fink, and Philipp Weiss (Paderborn, München: Fink, 2008), 39.

PART 2:

EXPOSING REALITY

CHAPTER THREE

THE "NEW CONQUERING EMPIRE OF LIGHT AND REASON": *THE CIVILIZING MISSION OF WILLIAM JONES*

NIHARIKA DINKAR

> But now all is to be changed. All the pleasing illusions, which made power gentle and obedience liberal, which harmonized the different shades of life, and which, by a bland assimilation, incorporated into politics the sentiments which beautify and soften private society, are to be dissolved by this new conquering empire of light and reason. All the decent drapery of life is to be rudely torn off. All the superadded ideas, furnished from the wardrobe of a moral imagination, which the heart owns, and the understanding ratifies, as necessary to cover the defects of our naked, shivering nature, and to raise it to dignity in our own estimation, are to be exploded as a ridiculous, absurd, and antiquated fashion.
> —Edmund Burke, *Reflections on the Revolution in France: And on the Proceedings in Certain Societies in London Relative to that event. In a Letter Intended to Have Been Sent to a Gentleman in Paris*

Written in the wake of the fall of the Bastille in France in 1789, Edmund Burke's politically charged pamphlet, *Reflections on the Revolution in France: And on the Proceedings in Certain Societies in London Relative to that event. In a Letter Intended to Have Been Sent to a Gentleman in Paris*, expressed his misgivings regarding the turn of events that had seen the French monarchy displaced and violent passions unleashed. It is, on the one hand, a lament for the "decent drapery" of a bygone era, but it saves its invective for the "barbarous philosophy" born of "cold hearts and muddy understandings" that functioned as a cabal and influenced opinion. While Burke directed his ire against this wider intellectual culture, his reference to the "new conquering empire of light and reason" was aimed at the central metaphor of Enlightenment thought, the empire of light and reason, and its vision of a naked truth.

What is interesting in Burke's characterization of light as a new conquering empire aligned with reason is that the intellectual foundation of Enlightenment thought is imagined not only in an enlightened discourse between philosophes that had broad public and historical resonance, but also includes a powerful visual dimension. Burke's protean phrase nods, on the one hand, towards the empire of light formed with the developments in artificial illumination that accompanied the industrialization of light, its instrumentalisation through an industry of representation (photography, film, magic lanterns, theatrical lighting, and so on), its new role in economic and industrial activity, its place in urban public life and spectacle, and its transformation of the home and individual interiors. Wolfgang Schivelbusch has described how the modern consciousness was forged amidst these dramatic transformations that sought to banish the night, so much so that according to Jonathan Crary, late capitalism ushers in an era of permanent illumination, a 24/7 where the distinction between night and day has been rendered irrelevant.[1] Sean Cubitt extends this notion to include a wider "practice of light" in modernity where visual media and its production of an aesthetics of dominance are seen as the latest episode in humanity's struggle to control light.[2]

On the other hand, Burke's conquering empire of light and reason also points to the political fact of empire, and of relationships between empire and colony. Through the course of the nineteenth century, light and darkness evolved into persistent metaphors that endorsed an ideology of progress between an imperial centre that dazzled with the lights of civilization and the primitive darkness of the peripheral colony. Tasked with bringing light into the benighted corners of the world, the civilizing mission of the British Empire cast its glare upon Indian epistemology, denouncing indigenous knowledge as superstitious darkness. Gayatri Spivak's '.notion of an epistemic violence cast this problem in the register of language and speech, where the erasure of local knowledge was viewed as an archival silencing of the subaltern voice.[3] Burke's reference to an empire of light and reason draws attention to the visual model at the heart of Enlightenment thought where light, truth, and reason are bound together in a powerful matrix with its promises of freedom. Tied to empire, it postulates a visual regime garnered around light and reason that emerged as a potent instrument of subjection in the colony.[4]

If the technologies, institutions, and practices that accompanied the industrialization of light granted a new texture to light, forming one crucial cog in an imperial vision machine, the veil functioned as another visual device that controlled light in an imperial optics. Although the iconography of unveiling had a long, hallowed tradition, it acquired new valence in its exposures of Oriental worlds that claimed to illuminate the darkness that lay behind the veil to offer a view into its hidden secrets, often aided by torches that marked a passage into the light. I propose that the unveiling of Oriental worlds should be viewed in conjunction with the simultaneous expansion of visual horizons that accompanied the scientific revolution, which presented optical technologies where the visible acquired a new legitimacy, bolstered by the extended invasive vision of these technologies.[5]

Pierre Hadot has identified the motif of unveiling nature as an antique trope with roots in the pre-Socratic idea that "nature loves to hide," proposing that it found renewed popularity in Enlightenment Europe as the "veil of Isis" and expressed the promises of science to expose the secrets of nature.[6] It was widely adopted by figures of the German Romantic tradition such as Schiller (*Das verschleierte Bild zu Sais*, 1795) and Novalis (*Die Lehrlinge zu Sais*, 1798–1799) and by artists, including William Hogarth in his *Boys Peeping at Nature* (1731) and Benjamin West in his *The Graces Unveiling Nature* (1779).

The veiled Isis figures as a prototype for a mysterious feminized Orient conflated with nature and the iconography of unveiling enacts the drama of sight in its quest to illuminate her secret interiorities. Remarking upon the currency of this feminized unveiling of Oriental mystique, Edward Said noted how "[t]he cultural, temporal and geographic distance [between the Occident and the Orient] was expressed in metaphors of depth, secrecy, and sexual promise: phrases like 'the veil of an Eastern Bride' or the 'inscrutable Orient' passed into the common language."[7] The secrets of a feminized Orient organized a visual economy of knowledge and concealment, where acts of unveiling mobilized a desire to see and to know, promising the disclosure of hidden secrets. *The Oriental Portfolio* (1839) for example, a lavishly produced travel portfolio featuring lithographs of India, featured a frontispiece of the graces unveiling an image of the Indian landscape, with subsequent pages leading the eye into the heartland and culminating in veiled interior spaces such as the harem, typically closed off to the outside eye. Such technologies of illuminating the darkness of the Indian landscape colluded with tropes of the civilizing mission, which sought to introduce the light of civilization into the "benighted darkness."

A marble commemorative statue of the Orientalist scholar William Jones (1799) by the sculptor John Bacon at St. Paul's Cathedral in London (fig. 3.1) presents an example of the how such iconographies of unveiling were conjoined with the civilizing mission. If the *Oriental Portfolio* had promised to uncover the darkness within which the Indian landscape had remained hidden to Western viewers and make visible its hidden interiorities, Bacon's sculpture heralded a triumphant moment that sought to introduce light into a benighted land. Jones stands in confident *contrapposto* wearing a toga that emphasizes his classical learning and holding in his hand a scroll, while leaning against a volume of *The Institutes of Manu*, an orthodox text on Hindu law that he had translated from the Sanskrit. On the pillar below, a set of books mark him out for his learning, callipers attest to his scientific spirit, a pair of scales recall his role as a judge of the Supreme Court at Calcutta, and a Ptolemaic globe and a lyre underline his aesthetic commitments and universal knowledge. Gazing out into the distance, he towers over a plaque at the base that describes his heroic conquest of Oriental knowledge.

On the base of the statue is a plaque with two allegorical male figures drawing back a drape on a roundel between them (fig. 3.2). The figure on the left holds aloft a torch in a gesture of triumph, the second figure leans over against the roundel with a lamp, and the metaphor of leading from darkness unto light is doubly articulated though devices of illumination as well as the drawing back of the curtain. Drawing back the drapes here specifically introduces light into a world of darkness. The Enlightenment trope of the light of reason dispelling the ignorance of darkness is appended to the liberatory rhetoric of the colonial civilizing mission here as the guardian figures holding the curtain perform a heroic act of rescue, ushering the pagan idolatry into the light. It is not accidental that the Indian figures in this image stand out sharply as "the much maligned monsters" characterized by Partha Mitter, set against the heroic Greco-Roman body of the colonizer.[8] The profusion of animal figures along with the zodiac ring and the "unnatural" image of the *trimurti* stand in for the arbitrary character of Indian belief sought to be "enlightened" by the towering figure of William Jones standing authoritatively above. Despite the centrality of the sculpture in the largest church in London, it has confounded art historians, with even contemporary sources deeming it fraught with cryptic symbolism.[9]

Like the *Oriental Portfolio* and its imagining of the nation as a two-dimensional image held aloft by allegorical figures in high relief, here too the Indian world is framed as picture in a roundel, long a symbol of possessive ownership (fig. 3.3). Bounded within the frame is a gamut of objects that claim to represent the nation – a sari-clad woman that one author has vaguely identified as Ganga, the "much maligned monstrous gods" that have been identified with Hinduism and zodiac imagery – a subject of much discussion in *Asiatick Research* (the journal of the Asiatic Society founded by Jones) as representative of Hindu chronology.[10] The imagery is culled from the *Puranas*, a body of literature that purports to represent the Indian historiographical tradition and would later come under attack by figures such as James Mill for its mythological character and irrational content. Bacon's sculpture, however, belongs to an earlier moment when an enthusiasm for Indology sought to recoup the Pauranik tales as independent corroboration of the biblical narrative.

Fig. 3.1: John Bacon, Memorial to William Jones, St. Paul's Cathedral, London, 1799.

Fig. 3.2: John Bacon, base of the memorial to William Jones, St. Paul's Cathedral, London, 1799

Fig. 3.3: John Bacon, detail of the base of the memorial to William Jones, St. Paul's Cathedral, London, 1799

An inscription above the roundel labels it "COURMA AVATAR," identifying the scene as that of the tortoise avatar of Vishnu. Here Vishnu in his tortoise guise intervened in the battle between the warring *devatas* (gods) and *asuras* (demons) as they churned the ocean in search of the nectar of immortality. During the course of the churning, using the serpent Vasuki as the rope and Mount Mandara as the staff, the mountain began to sink. Vishnu, in his tortoise avatar, performed a heroic act of rescue as he held up the mountain upon his carapace. The churning oceans threw up fourteen treasures, including the sun and the moon, Indra's elephant, the horse that draws the chariot of the sun god, and the bull, all of which are represented around the central four-armed figure of Vishnu atop the churning staff. In his essay "The Gods of Greece, Italy and India," William Jones interpreted the cosmic floods of the tortoise avatar of Vishnu as an allegorical rendering of the deluge described in the Old Testament, and hence the Pauranik narrative served as an affirmation of the biblical narrative.[11] Thomas Maurice (1754–1824), Oriental scholar and keeper of manuscripts at the British Museum, corroborated this reading of the tortoise avatar. He found in the rainbow, one of the treasures thrown up by the churning ocean, a similarity with Noah's story where the rainbow appears at the end of the Flood.[12]

Bacon's source for the roundel appears to be an engraving from Thomas Maurice's *The History of Hindostan* (1795), which included a representation of Vishnu in the guise of a tortoise, with the inscription noting his act of rescue from the deluge (fig. 3.4).[13] Maurice himself claimed that he had copied the illustration from a volume in the possession of the Society of Antiquaries that listed the nine avatars of Vishnu. Bacon's copy is almost identical to Maurice's, with the horned *asuras*, the placement of the elephant and the horse, and even the churning staff itself. In *The History of Hindostan*, Maurice had set himself the task of writing an antediluvian history that contained, (in the extended title of the book) "Indian and other Oriental Accounts of the General Deluge," while also discussing Indian cosmogony, astronomy, and the solar and lunar zodiacs of Asia, among others. His account of the "Courma Avatar" is followed by an analysis of its astronomical allusions that relies largely on William Jones's proposition that the tortoise avatar related to Hindu astronomy. In each case, the goal was to seek a convergence between Hindu and Mosaic mythologies that recent scholarship has interpreted as a means of assimilating the strangeness of Hinduism in the late eighteenth century, so much so that pagan idolatry has found a way into the central dome of the most significant church in London.[14]

Fig. 3.4: "Courma Avatar", plate 9 from Thomas Maurice, *History of Hindostan, Vol. 1* (London: W. Bulmer and Co., 1795)

One of the most predominant myths about the convergence between Hindu and Mosaic ideologies related to the *trimurti*. In Bacon's roundel, the *trimurti* that stands to the left of the sari-clad figure undoubtedly recalls the colossal three-faced figure of Siva from Elephanta, given the amount of attention the sculpture had received in antiquarian circles.

While more recent scholarship has identified the figure as Mahesamurti, available accounts in the late eighteenth century, for example those of the German cartographer and explorer Carsten Niehbuhr, explained it as a composite figure of Brahma, Vishnu, and Siva. Thomas Maurice invoked both William Jones and Niehbuhr to identify the figure as *trimurti* and relate it to the Christian trinity, finding further affinities between the Brahmanical religion and rites and those of the Druids in his own country. It is unclear which visual sources Bacon drew upon, given that his version features decidedly conical headgear (*karand mukut*) rather than the flattened version (*kirit mukut*) of Niehbuhr. Considering the fact that the influential iconographic resource, Edward Moor's *Hindu Pantheon*, was only published in 1810, James Forbes' drawings of the altarpiece at Elephanta (1775), which depict pointed headgear, probably served as a more likely source.

Maurice emerged as a keen interpreter of Jones's writings and Bacon's reliance upon him is understandable not only because of the esoteric nature of Jones's work, but also because Bacon shared his deep-seated religious beliefs and a common patron, the East India Company. Maurice was a clergyman by profession whose first writings on India were written against the backdrop of the French Revolution and "neither his conviction, the result of education and reflection, nor his profession, would permit him to publish anything respecting India without an effort at least to refute the argument and subvert the hypothesis of the atheists of the day, who had taken their stand to endeavour to root out Christianity and demoralise the world."[15] He was, moreover, an ardent admirer of William Jones, whom he had met during his early days at Oxford and upon whose death he had penned an elegiac poem. In his preface to the *History of Hindostan*, he claimed "perfect co-incidence with the opinion of Sir William Jones," relying upon his authoritative scholarship for his popular histories of India. His writings were in fact widely embraced, as evidenced by the fact that Samuel Taylor Coleridge copied a passage from Maurice's accounts of the icy caves of Kashmir (probably Amarnath) from *History of Hindostan* for his definitive Romantic poem, *Kubla Khan* (1816).[16]

Bacon himself was a devout Christian and founding member of the Eclectic Society (later the Church Missionary Society) who likely sympathized with Maurice's views about Orientalist mythologies confirming the truth of the Bible. His biographies all underline his piety: "Religion with him was not the Sunday-coat of a formalist . . . but a change of heart and a hope full of immortality, grounded alone on the work of a Redeemer."[17] Bacon's interpretation of Jones's contributions to Indian literature and culture appears to be closely aligned with official

East India Company policy. During the 1790s, Bacon developed a close relationship with the East India Company, having contributed a statue of the war hero Cornwallis, as well an allegorical sculptural frieze for the pediment outside the East India House. At this time, he also acquired three large commissions for St. Paul's Cathedral, the last one being that of William Jones. When his design for the monument of William Jones was approved in 1796, the Secretary of the East India Company noted in his letter to Bacon their agreement over the 'Ideas and Sentiment to be expressed.'[18] His allegorical style, with a predilection for classical tropes, found favour with his East India patrons despite his lack of training in Rome. In his biographies, Bacon emerges as a resolute nationalist and devout Christian whose lack of Continental training is dismissed to recall pride in a home-grown education.[19]

The zodiac ring that crowns the roundel was testimony to the great interest in astronomy not only on the part of Oriental scholars who sought to understand Hindu cosmogony, but also by Company officials for whom astronomy was an efficient means to establish the latitude and longitude of sites for conquest by the army, and both these sets of studies were published in *Asiatick Research,* the journal of the Asiatic Society.[20]

Hindu cosmogony straddled a world between astronomy, mathematics, and divination and engaged with both Pauranik notions of the cosmos where the earth was considered a flat disk in an egg shaped universe as well as Siddhantic models of Earth as a fixed sphere around which the sun, moon, and planets revolved.[21] William Jones had taken an active interest in Hindu astronomy and published three papers in *Asiatick Research* between 1790 and 1792, and was accompanied by others, including John Bentley and Samuel Davis. Thomas Maurice, for his part, made astronomy the centrepiece of his scholarship, explaining his interest as "mak[ing] that exalted science subservient to nobler purposes; to collect in one centre the blended rays shed by the heavenly orbs, and direct their powerful focal splendour to the illustration of those grand primeval truths which form the basis of the national Theology; a Theology so inseparably connected with the National Government"[22]

Maurice's invocation of heavenly light to reveal the primeval truth is a particularly apt metaphor that ties together both Bacon's symbolism of the revelatory powers of light and Jones's own interpretation of its centrality in the Sanskrit text he had translated, as he notes in the preface:

> The many panegyricks on the *Gayatri*, the *Mother*, as it is called, of the Veda, prove the author to have *adored* (not the visible material *Sun*, but) that *divine and incomparably greater light*, to use the words of the most venerable text in the *Indian* scripture, *which illumines all, delights all, from*

which all proceed, to which all must return, and which alone can irradiate (not our visual organs merely, but our souls and) *our intellects.*

Jones attributes to Manu a dismissal of the heavenly light of the sun for the greater light of knowledge, a metaphor that finds easy acknowledgement within Enlightenment thought. However, this does not cast the Indian lands as 'enlightened'. The laws of Manu remain in fact "a system of despotism and priestcraft, both indeed limited by law but artfully conspiring to give mutual support, though with mutual checks; it is filled with strange conceits in metaphysics and natural philosophy, with idle superstitions, and with a scheme of theology most obscurely figurative, and consequently liable to dangerous misconception . . ."

Jones therefore feels compelled to introduce light into this benighted world.The conquest of the irrationality of Hindu law by an "enlightened" Jones, bringing it within the fold of the colonial administrative machinery, marks the passage from darkness unto light. Jones's act of translating the Manu-Smriti was undertaken with the noble mission of administering justice in accordance with native laws, but it also displaced Sanskrit scholarship in legitimizing an authoritative version. The codification of the polymorphous system of legal codes into an Ur-text that could ultimately prove useful as instruments of governance allowed Jones to override internal disputes among native specialists. The alliance with the law is therefore not accidental, its aim being to establish a moral and legal code within which claims for justice are addressed.[23] Gayatri Spivak has in fact referred to the colonial codification of Hindu law (by Jones and other figures, including Charles Wilkins and Henry Colebrooke) as the very epitome of epistemic violence representing, as it did, an effective means of control.[24]

Even as enthusiastic Indological scholars translated and made available an entire body of Sanskrit texts to a Western audience, there was an unspoken understanding that they were rescuing the literature of an ancient past that threatened to be lost forever. Implicit in this exercise was the notion that the British were bequeathing to the Indians the Indians' own history, which had remained inaccessible to them through the corrupted present, and that the "enlightened" British were recouping. Bacon's monumental sculpture of the towering Jones clad in classical garb encapsulates this relationship of knowledge and power, with the trophy-like roundel featuring Pauranik gods ushered into the light of civilization by torch-bearing Greco-Roman guardian figures. A contemporary monument to William Jones by John Flaxman at the University College Chapel at Oxford features a somewhat more subtle expression of the relationships between Jones and native knowledge. Jones is seated on an elevated bench composing his digest on a desk facing three seated native men on a lower platform. A banana tree sits against the background, while the inscription underneath it underscores his accomplishment, "He formed a digest of Hindu and Mohammedan laws." While the seated Brahmins were presumably the scholars with whose help Jones compiled his digest, the relationship is reversed, presenting Jones as a tutor schooling his pupils. More obvious examples of the paternalistic schooling of childlike Indian subjects by British guardian figures is included in the monuments of missionaries, such as the statue of Bishop Middleton at St. Paul's Cathedral (London), in which the bishop is depicted as blessing two kneeling Indian subjects, or Francis Chanterey's monument to Reverend Reginald Heber in St. George's Cathedral in Madras, which features the same iconography. The language of leading from darkness unto light takes on another set of meanings in the missionary vocabulary, but its similarity to Bacon's rendition underlines the continuities between the two.[25]

Despite the honour accorded to Jones for his pioneering work, his digest of laws ultimately had little impact upon the Indian legal code, and it fell into disuse after 1864. It has been noted that Jones's translation had a greater influence on Oriental scholarship than within the colonial administration itself.[26] Its triumph lay in the deciphering of Oriental mystique, an aspect uncannily grasped in Bacon's sculpture, which pitted the arcane symbolism of the Pauranik deities against the towering presence of Jones and the heroic Greco-Roman bodies of the guardian figures. Although recent accounts have expressed puzzlement over its "cryptic" symbolism, Bacon's design was not only cognizant of popular Oriental scholarship but also seemingly in step with what Edmund Burke referred to as the "new conquering empire of light and reason" in his remarks on the revolution in France.[27] In its evocation of key metaphors of Enlightenment ideology, Burke's reflections lamented the passing of the old order that left "all the decent drapery of life . . . rudely torn off."[28] This viewpoint also reflected in a series of frontispieces in seminal books by Enlightenment philosophes that invoked the metaphor of the light of reason, among them Andrew Motte's 1729 translation of Newton's *Principia*, Voltaire's *Elements of the Philosophy of Newton* (1738), and, most famously, Charles Nicholas Cochin's drawing for Diderot and D'Alembert's *Encyclopédie* (1772).[29] In appending the allegory of light and reason to the political circumstance of empire, Bacon did no more than make visible the visual regime, which drew empire and colony into a network of relations between vision, knowledge, and power.

Notes

[1] Wolfgang Schivelbusch, *Disenchanted Night: The Industrialization of Light in the Nineteenth Century*, trans. Angela Davis (Berkeley: University of California Press, 1988). See also Frances Guerin, *A Culture of Light* (Minneapolis, London: University of Minnesota Press, 2005); Jonathan Crary, *24/7: Late Capitalism and the Ends of Sleep* (London and New York: Verso, 2013).

[2] Sean Cubitt, *A Practice of Light: A Genealogy of Visual Technologies from Prints to Pixels* (Cambridge, MA: MIT Press, 2014)

[3] Gayatri Chakravorty Spivak, "Can the Subaltern Speak?," in *Can the Subaltern Speak? Reflections on the History of an Idea*, ed. Rosalind Morris, (New York: Columbia University Press, 2010), 21–78.

[4] On the tyranny of imperial reason, see Gyan Prakash, *Another Reason: Science and the Imagination of Modern India* (Princeton: Princeton University Press, 1999)

[5] Barbara Stafford presents this argument in *Body Criticism: Imaging the Unseen in Enlightenment Art and Medicine* (Cambridge, Mass.: MIT Press, 1991).

[6] Pierre Hadot, *The Veil of Isis: An Essay on the History of the Idea of Nature* (Cambridge, MA: Harvard University Press, 2006)

[7] Edward Said, *Orientalism* (New York: Vintage Books, 1994), 222.

[8] Partha Mitter, *Much Maligned Monsters: A History of European Reactions to Indian Art* (Oxford: Clarendon Press, 1977). Given the Greco-Roman characterization of the sculpture, the significance of the roundels as serving within an imperialistic metaphor of annexation and display as trophy deserves to be noted.

[9] Barbara Groseclose, *British Sculpture and the Company Raj*, p. 103.

[10] Barbara Groseclose proposes the draped figure is 'mother Ganges' given its affinity with a similar figure at the base of John Bacon Jr.'s monument to James Achilles Kirkpatrick. *British Sculpture and the Company Raj*, p. 102–3.

[11] William Jones, "On the Gods of Greece, Italy and India" in *The Works of Sir William Jones: With the Life of the Author*, ed., Lord Teignmouth, vol. 3 (London: Printed for John Stockdale and John Walker, 1807), 338.

[12] Thomas Maurice, *Sanscreet Fragments, or Interesting Extracts from the Sacred Books of the Brahmins, on Subjects Important to the British Isles. In two parts.* (London: Thomas Maurice, 1797), 20.

[13] Maurice notes that the engravings of Vishnu's avatars (he includes the Varaha avatar and the Matsya avatar) were "fac similes" from painted versions in temples and hence utterly oblivious to the rules of perspective. If they found admirers it was only because of their eccentric originality, he claimed. Thomas Maurice, *History of Hindostan,* vol. 1 (London: W. Bulmer and Co., 1795), 40.

[14] Thomas Trautmann, *Aryans and British India* (New Delhi: Yoda Press, 2004), 74–80.

[15] Obituary of Rev. Thomas Maurice, *Gentleman's Magazine* 94: 1, Dec.– Jun. 1824: 469. His obituary also claims he is descended from the princes of Powis in Wales. Powis Castle was the site of a major collection of Indian art and antiquities spirited away by Robert Clive after the Battle of Plassey in 1757 when he returned to Britain with a personal fortune of 234,000 pounds, making him the richest self-

made man in Europe. After Clive's son married Lord Powis's daughter in 1784, the estates merged in 1801.

[16] Deirdre Coleman, "The 'Dark Tide of Time': Coleridge and William Hodges' India" in *Coleridge, Romanticism and the Orient: Cultural Negotiations*, ed. David Vallins, Kaz Oishi, Seamus Perry (London: Bloomsbury, 2013), 39–54.

[17] Allan Cunningham, *The Lives of the Most Eminent British Painters and Sculptors*, vol. 3. (New York: Harper & Brothers, 1846), 194.

[18] Cited in Ann Cox-Johnson, *John Bacon, R. A. 1740–1799* (London: St. Marylebone Society Publication, 1961), 38.

[19] See ibid., and Richard Cecil, *Memoirs of John Bacon Esq. R.A* (London: F & C Rivington, 1801).

[20] Srabani Sen, "The Asiatic Society and the Sciences in India, 1784–1947," in *History of Science, Philosophy and Culture in Indian Civilization*, vol. 15, part 4, *Science and Modern India: An Institutional History*, ed., Uma Das Gupta, (New Delhi: Centre for Studies in Civilizations, 2011), 27–68.

[21] See Christophe C. Minkowski on how the two were reconciled. "The Pundit as Public Intellectual: The Controversy over *Virodh* or Inconsistency in the Astronomical Sciences" in *The Pundit: Traditional Scholarship in India*, ed. Alex Michaels (New Delhi: Manohar, 2001), 79–98.

[22] Thomas Maurice, *History of Hindostan,* xxii

[23] For a closer analysis of the metaphor of light in Jones' translation project see Jenny Sharpe, "The Violence of Light in the Land of Desire; Or, How William Jones Discovered India", *boundary 2* 20, no. 1 (1993): 26–46.

[24] Gayatri Spivak, "Can the Subaltern Speak?"in *Marxism and the Interpretation of Culture*, ed. Cary Nelson and Lawrence Grossberg (Urbana: University of Illinois Press, 1988), 281.

[25] Despite his invocations of the light of knowledge, Jones posits Christianity as the only true revelation, an attitude that resonates with Maurice and Bacon. For an explicit account of Jones' religious beliefs and their impact on intellectual history see Martin Priestman, *Romantic Atheism: Poetry and Freethought, 1780–1830* (Cambridge: Cambridge University Press, 2004), 44–79. In his preface to the Institutes, Jones writes: "Whatever opinion in short may be formed of MENU and his laws, in a country happily enlightened by sound philosophy and the only true revelation, it must be remembered, that those laws are actually revered, as the word of the Most High, by nations of great importance to the political and commercial interest of Europe, and particularly by many millions of Hindu subjects, whose well-directed industry would add largely to the wealth of Britain, and who ask no more in return than protection for their persons and places of abode, justice in their temporal concerns, indulgence to the prejudices of their old religion, and the benefit of those laws, which they have been taught to believe sacred, and which alone they can possibly comprehend."

[26] J. Duncan Derrett, *Religion, Law and the State in India* (New York: The Free Press, 1968) 250, cited in Jenny Sharpe, "The Violence of Light in the Land of Desire; Or, How William Jones Discovered India", *boundary 2* 20, no. 1 (1993): 26–46.

[27] Recent accounts that have discussed Bacon's sculpture and expressed a baffled curiosity include Barbara Groseclose, *British Sculpture and the Company Raj* (Newark, DE: University of Delaware Press, 1995), 103, Nicholas Penny, *Romantic Church Monuments* (New Haven, CT and London: Yale University Press, 1977), 146. Jason Edwards has pointed to the problems in deciphering the obscure imagery in "Introduction: From the East India Company to the West Indies and Beyond: The World of British Sculpture, c. 1757–1947," *Visual Culture in Britain* 11, no.2 (2010): 147–172.

[28] Edmund Burke, *Reflections on the Revolution in France*, ed. J.G.A. Pocock, (Indianapolis: Hackett Publishing, 1987), 66–67.

[29] See James Schmidt, "This New Conquering Empire of Light and Reason: Edmund Burke, James Gillray, and the Dangers of Enlightenment," *Diametros* 40 (2014): 126–148.

CHAPTER FOUR

"LITTLE POINTS OF BRILLIANCE DANCING ON GREY STONES": LIGHT AND SHADOW IN GUSTAVE FLAUBERT'S *MADAME BOVARY* AND IN IMPRESSIONIST ART

MERAV FIMA

Although Gustave Flaubert's 1857 novel *Madame Bovary* has been classified by many scholars as realist literature, it has much in common with impressionism, the artistic style that developed in the visual arts several years later. Indeed, Flaubert's literary endeavour is compatible with Jesse Matz's definition of literary impressionism in his book *Literary Impressionism and Modernist Aesthetics*, where he states that: "pictorial descriptions of shifting light and colour, subjective accounts of sensuous experience, transmission of immediate and evanescent feeling – these are literary Impressionism's specialties."[1]

Flaubert (1821–1880) himself argued against scholars who classified *Madame Bovary* as a realist novel, claiming that "it was in hatred of realism that I undertook this book."[2] Instead of depicting life as it really was, Flaubert strove to aestheticize bourgeois existence. He declared "I wanted to show that bourgeois dreariness and mediocre sentiments could sustain beautiful language."[3] This statement points to Flaubert's emphasis on the formal qualities of the novel rather than its content. In his book *The Rules of Art*, Pierre Bourdieu links Flaubert's literary endeavour to painter Edouard Manet's innovations in the visual arts, which bridged realism and impressionism. He claims that both artists strove for a "pure aesthetic" while continuing to represent realistic situations.[4] Bourdieu writes that "[Flaubert's] mastery of all the requirements of form allows him to assert almost limitlessly the power he possesses to constitute aesthetically any reality in the world, including those which, historically, realism had taken as its chosen objects."[5]

Instead of depicting heroic subject matter, which had been the focus of most great artworks until then, Flaubert chose to represent the common life of the bourgeoisie, showing it at its worst, as in Emma's adulterous relationships with Rodolphe and Léon. A similar focus on the bourgeoisie and their modern leisure activities is apparent in impressionist paintings. Like Flaubert, these artists adopted banal, contemporary subject matters, but created beautiful artworks through their composition, artistic technique, and the materials they used.[6]

Despite the fact that the publication of this novel preceded the first impressionist exhibition by seventeen years,[7] Flaubert seems to have embraced many elements of the impressionist aesthetic even before its official establishment. Most notably, Flaubert creates special optical effects through his use of light and colour in a way that resembles the impressionist painters' application of paint to the canvas in order to capture the light's play on objects, a technique that imbues their works with luminosity.

According to Adrianne Tooke, Flaubert uses light, shadow, and colour to evoke images in the reader's mind, although his "images are . . . reduced to glimpses and fragments, colours and shapes, contrasts of light and dark . . . reflecting his leitmotif of paintings half obscured by time, light, shadow, and other things, and becoming, paradoxically, all the more effective for that."[8] Flaubert's description of the Cathedral of Rouen in sunlight, as seen by Léon on his way to meet Emma, evokes Claude Monet's paintings of the Cathedral of Rouen. When Léon approaches the cathedral, he sees the facade glimmering in the sunlight: "It was a fine summer morning . . . the light sloping down on the Cathedral set little points of brilliance dancing on the grey stones. A flock of birds wheeled in the blue sky, round the trefoiled bell-turrets."[9]

Between 1892 and 1894, Claude Monet painted a series of thirty paintings of the western facade of the Cathedral of Rouen.[10] Each painting is unique, since the light hits the building differently under diverse meteorological conditions, at different times of day, and in different seasons.[11] In Monet's paintings, as in Flaubert's description, the Rouen Cathedral appears to shimmer in broad daylight. The light exerts such a strong effect on the building that the bright colours seem more substantial than the structure itself, which almost dissolves. More precisely, "the myriad details of a Gothic cathedral dissolve into light and shadow, which are indicated by individual patches of colour."[12] As in Flaubert's description, the yellow sunlight hitting the facade does appear to "dance" on the stones, contrasting with the blue sky. The visual effect is intensified due to the fact that blue and yellow are complementary colours that accentuate each other when placed side by side, according to the RGB colour model, which applies to colours created by light. The impressionist artists, too, juxtaposed complementary colours to create a vibrant visual impression, based on the scientific theory expounded by Michel Chevreul. Flaubert, therefore, achieves in words what Monet would later accomplish in paint.

Flaubert's attention to the effects of different kinds of light is apparent in the scene at the Rouen Cathedral. He writes "Broad daylight shone in through the three open doors and stretched down the whole length of the Cathedral in three enormous rays [...] The crystal chandeliers hung motionless. In the choir a silver lamp was burning."[13] The contrast between natural and artificial light was of great interest to the impressionist artists and is echoed in this passage, where sunlight is juxtaposed with the light cast by the chandeliers and the lamp. A similar effect is created earlier in the novel during Emma's visit to the church in Yonville-l'Abbaye, where she seeks out the priest and requests a spiritual remedy for all that ails her. The church is illuminated thus: "At the far end of the church a lamp – a nightlight wick inside a hanging glass – was burning: it looked from a distance like a little white cloud quivering above the oil. A long ray of sunlight shone down the nave, making the aisles and the recesses still darker."[14] This passage depicts an interesting optical effect caused by the juxtaposition of natural and artificial light, with the surprising result that the aisles seem even darker in contrast to the bright sunlight shining down the nave.

Impressionist artists were also interested in the play of light and shadow, rendering the shadow in the complementary colour (red/green, blue/orange, yellow/purple) and giving it as much prominence as the concrete object.[15] Flaubert, too, explores this notion in the passage in which Léon fantasizes about Emma's arrival at the Cathedral, placing the lovers in the shadow of the arches, while illuminating Emma's face with light coming in through the windows: "The Cathedral was like a gigantic boudoir prepared for her. The arches leaned down into the shadows to catch her confession of love, the windows shone resplendent to light her face, and the censers would burn that she might appear as an angel in an aromatic cloud."[16] Here the lighting creates a sensual atmosphere befitting the illicit act anticipated by Léon.

Flaubert also uses the evening at the opera as an opportunity to examine the effects of different types of indoor lighting on the characters. Just before the beginning of the performance, "the orchestra candles were lit and the chandelier was let down from the ceiling, its glittering facets shedding a sudden gaiety upon the house."[17] The gaiety experienced by the audience reflects the physical and emotional sensations caused by the change in lighting. Flaubert's description thus captures both the appearance of the artificial light and its effect on the novel's characters. This interpretation is supported by Rosemary Lloyd, who claims that light is used in this novel to reflect the characters' emotions: "Different forms of light can be an essential element of the emotional landscapes," she writes. "The importance of light is not merely to create a 'realistic' representation of the world in which Emma finds herself, but also to suggest emotion."[18] Judith Poole Knapp takes this hypothesis even further, suggesting that light is symbolic of the characters' psychological state. She writes: "Colour and light . . . serve as vehicles for Flaubert's characteristic use of symbolism. When taken cumulatively throughout the novel, the meanings ascribed to certain colour and lighting effects often symbolize specific situations or a character's psychology, while at the same time reflecting a particular point of view."[19]

In Jean Béraud's 1877 painting *The Lobby of the Paris Opera*, the lit lamps and chandeliers feature prominently, in sharp contrast to the dark shadows cast by the imposing arches, creating an expectant atmosphere. The opera and the theatre were among the main activities enjoyed by the French bourgeoisie in the mid-nineteenth century. Flaubert and several impressionist painters captured the experience of the opera; most of their paintings show very little of the performance, depicting, rather, audience members observing other guests through their binoculars.[20] Likewise, in *Madame Bovary*, much space is taken up by descriptions of people attending the opera. Flaubert writes:

> [Emma's] heart beat as she entered the foyer. She smiled involuntarily, out of vanity, as she saw people jostling down the passage on the right while she went up to the boxes . . . when she was ensconced in her box she drew herself up in her seat as to the manner born. The theatre was beginning to fill up. Opera-glasses emerged from their cases; seat-holders, catching sight of their friends, waved greetings. They had come to find relaxation in the arts from the cares of business.[21]

Mary Cassatt's painting *Woman in Black at the Opera* of 1879 seems to illustrate this very scene.[22] The woman observing the other audience members through her binoculars appears to be ensconced in the shadow of the box, while a bright artificial light illuminates the glowing golden balconies. As in impressionist paintings of the opera that depict the audience rather than the performance, Flaubert minimizes the performance, representing instead a "slice of life."

Focusing on the production of beauty in art, Flaubert depicted common bourgeois pursuits with great technical skill, instead of representing heroic and aristocratic subject matters. Moreover, many of his descriptions, which at first seem objective and naturalistic, are rendered from the subjective point of view of the narrator or of one of the characters, who shares his own impressions of the object with the reader. While the Impressionist artists sought to depict a "slice of life" as it appeared objectively, they, too, were mainly concerned with capturing their own sensations as evoked by the visual experience, creating "an art of pure sensation."[23]

In their introduction to *Artistic Relations*, Peter Collier and Robert Lethbridge assert that Flaubert is among the French writers who have "tried to provide [their] own apparatus for visualization, inviting each reader to create a private and unrepeatable inner vision, through a largely Impressionistic style of writing."[24] Alan Raitt reiterates this idea in his book *The Originality of Madame Bovary*, stating that "Flaubert has no intention of providing the plethora of information which would make it possible for the reader to visualise a character or a place exactly as he himself visualises it. In that way, it would be legitimate to call his descriptions impressionistic rather than naturalistic."[25] In one of Charles's first encounters with Emma, she is described thus, obviously from the beholder's subjective point of view: "She opened [the parasol] out; it was of shot silk, and the sun shining through it cast flickering lights over the white skin of her face. She smiled in the moist warmth beneath it, and they heard the drops of water dripping on to the taut silk one by one."[26] Emma is described from Charles's subjective point of view as he falls in love with her and is captivated by her charms, while his perception of her is coloured by the light filtering through the silk of the parasol. This passage is evocative of Claude Monet's *Study of a Figure in Nature: Woman with an Umbrella Turned to the Right* from 1886, which also features a woman holding a parasol, her indistinct facial features blurred by the white light scintillating against it, represented by short, broken brushstrokes.[27] In both cases, the figures clearly constitute the character's or painter's perceptual impression, rather than an objective portrayal.

Descriptions of this kind may be associated with Edmond de Goncourt's concept of *écriture artiste* (artistic writing), which combines realistic writing with techniques borrowed from the visual arts, specifically impressionism. Here language is employed in an artistic manner, making an ordinary encounter seem extraordinary through its impressionistic description, which gives us a general, yet imprecise, sense of Emma's appearance.[28] Indeed, the impressionists sought to capture transitory moments, as manifested by fleeting effects of light and atmosphere. As Laurence Porter and Eugene Gray assert, "Flaubert concentrates on presenting the impressions experienced by his characters . . . These impressions frequently represent a psychological reaction to a scene or event."[29] Hence, Flaubert's incorporation of the characters' feelings, impressions, and sensations into his descriptions of other characters and events is a strong indication of his impressionistic tendencies.

Like Flaubert, impressionist artists adhered to the tenets of the "art for art's sake" movement, ignoring their social role and focusing solely on the production of beauty in art. More precisely, with the rise of impressionism, the material, medium, form, and style became more important than the often mundane content of the artwork. The American impressionist painter James Abbott McNeill Whistler famously defended his artistic style in a libel case against John Ruskin, who accused him of "flinging a pot of paint in the public's face," by claiming that he believed in "art for art's sake" and that his paintings examine the effects of colour and light, as well as experiment with form, and are not meant to reflect reality.[30]

The following citation from *Madame Bovary* is reminiscent of the very approach described above by Whistler:

> Full and flushed, the moon came up over the skyline behind the meadow, climbed rapidly between the branches of the poplars, which covered it here and there like a torn black curtain, rose dazzling white in the clear sky, and then, sailing more slowly, cast down upon the river a great splash of light that broke into a million stars, a silver sheen that seemed to twist its way to the bottom, like a headless snake with luminous scales; or like some monstrous candelabra dripping molten diamonds. The soft night was all about them. Curtains of shadow hung amid the leaves.[31]

This passage immediately follows Rodolphe's assertion of his adoration, which Emma forcefully extracts from him, on the evening preceding their planned elopement from Yonville-l'Abbaye. The expressions of the different qualities of the moonlight shimmering against the water paint a vivid picture in the reader's mind.

Like Whistler's paintings from the "Nocturne" series,[32] the above description constitutes "a representation of moonlight. My whole scheme was only to bring about a certain harmony of colour . . . It's an artistic arrangement." Whistler goes on to say: "By using the word 'nocturne' I wished to indicate an artistic interest alone . . . A nocturne is an arrangement of line, form, and colour first."[33] Flaubert would certainly have sympathized with this view, since he also worked hard to perfect his artistic form and style in order to beautify a banal story.[34]

The poet Leconte de Lisle declared that Flaubert was "a poet and a painter like few others,"[35] implying a strong connection between his writing and the visual arts. Indeed, this praise is well-deserved, since his novels, particularly *Madame Bovary*, address the same issues and employ the same artistic techniques as impressionist paintings. Specifically, *Madame Bovary* focuses on the mundane activities of the bourgeoisie, but excels in its form; records the impressions and reactions of the narrator and characters to certain objects, people, or events; and explores the effects of light and colour. Thanks to these innovative artistic techniques, Flaubert's novel and the Impressionist paintings that followed it were revolutionary at the time of their creation and have had a lasting, illuminating effect on modern Western culture.

Notes

[1] Jesse Matz, *Literary Impressionism and Modernist Aesthetics* (Cambridge: Cambridge University Press, 2001), 3.

[2] Letter to Madame Roger des Genettes, October or November 1856, in Gustave Flaubert, *Correspondance*, ed. Eugène Pasquelle (Paris, 1900), 4:134, quoted in Pierre Bourdieu, *The Rules of Art: Genesis and Structure of the Literary Field*, trans. Susan Emanuel (Stanford: Stanford University Press, 1995), 92.

[3] Quoted in Bourdieu, *The Rules of Art*, 93.

[4] Ibid., 105–106.

[5] Ibid., 108.

[6] Laurie Schneider Adams, *Art across Time*, 2nd ed. (New York: McGraw-Hill, 2002), 805.

[7] The first impressionist exhibition was held by the Anonymous Society of Painters, Sculptors, Engravers, etc., at the photographer Nadar's studio at 35 Boulevard des Capucines, Paris, from April 15 to May 15, 1874.

[8] Adrianne Tooke, *Flaubert and the Pictorial Arts: From Image to Text* (Oxford: Oxford University Press, 2000), 239.

[9] Gustave Flaubert, *Madame Bovary: A Story of Provincial Life*, trans. Alan Russell (Middlesex: Penguin Books Ltd., 1971), 250.

[10] See, for example: Claude Monet (1840–1926), *Rouen Cathedral, the West Portal and Saint-Romain Tower, Full Sunlight, Harmony in Blue and Gold*, 1894, oil on canvas, 107 x 73 cm, Musée d'Orsay, Paris.

[11] Gloria K. Fiero, *The Humanistic Tradition*, 3rd ed., vol. 5 (New York: McGraw-Hill, 1995), 116.

[12] Schneider Adams, *Art across Time*, 826.

[13] Flaubert, *Madame Bovary*, 251.

[14] Ibid., 124.

[15] Schneider Adams, *Art across Time*, 805.
[16] Flaubert, *Madame Bovary*, 251.
[17] Ibid., 233–234.
[18] Rosemary Lloyd, *Madame Bovary* (New York: Routledge, 1990), 121–122.
[19] Judith Poole Knapp, "Madame Bovary: The Dialectics of Color and Light" (PhD diss., University of Arizona, 1980).
[20] See, for example: Pierre-Auguste Renoir (1841–1919), *The Theatre Box*, 1874, oil on canvas, 80 x 63.5 cm, Courtauld Institute Galleries, London.
[21] Flaubert, *Madame Bovary*, 232–233.
[22] See Mary Cassatt (1844–1926), *Woman in Black at the Opera*, 1879, oil on canvas, 81.28 x 66.04 cm, Boston, Museum of Fine Arts (Charles Henry Hayden Fund).
[23] Fiero, *Humanistic Tradition*, 113.
[24] Peter Collier and Robert Lethbridge, eds., *Artistic Relations: Literature and the Visual Arts in Nineteenth-Century France* (New Haven: Yale University Press, 1994), 10.
[25] Alan Raitt, *The Originality of Madame Bovary* (New York: Peter Lang Press, 2002), 71.
[26] Flaubert, *Madame Bovary*, 30.
[27] See Claude Monet (1840–1926), *Study of a Figure in Nature: Woman with an Umbrella Turned to the Right*, 1886, oil on canvas, 130.5 × 89.3 cm, Musée d'Orsay, Paris.
[28] Richard Hobbs, "Visual Display in the Realist Novel: 'l'aventure du style'," in *Impressions of French Modernity*, ed. Richard Hobbs (Manchester: Manchester University Press, 1998), 120–125.
[29] Laurence M. Porter and Eugene F. Gray, eds., *Gustave Flaubert's Madame Bovary: A Reference Guide* (London: Greenwood Press, 2002), 79.
[30] Schneider Adams, *Art across Time*, 835–836.
[31] Flaubert, *Madame Bovary*, 210.
[32] The title of the series, "Nocturne," is a musical term. See, for example, James Abbott McNeill Whistler, 1834–1903, *Nocturne in Grey and Gold – Canal*, 1884, oil on canvas, Freer Gallery of Art, Washington, DC.
[33] Quoted in Richard Dorment and Margaret F. MacDonald, eds., *James McNeill Whistler* (New York: Harry N Abrams, 1995), 122.
[34] Porter and Gray, *Gustave Flaubert's Madame Bovary*, 70–74.
[35] "Tu es un poète et un peintre comme il y en a peu." Letter by Leconte de Lisle, December 16, 1862, in *Gustave Flaubert: Oeuvres complètes et Annexes,* 1221. Quoted in: Tooke, *Flaubert and the Pictorial Arts*, 1.

CHAPTER FIVE

REDISCOVERING LUCAS CRANACH THE ELDER BY MEANS OF LIGHT-BASED TECHNOLOGY

RUTH LUBASHEVSKY-AMAR

Authentication and Attribution

In May 2000, Christie's and Sotheby's released their spring catalogues only to discover that they were both offering Paul Gauguin's *Vase de Fleurs (Lilas).*The two paintings were sent to an expert, who determined that the one held by Christie's was the forgery. Sotheby's sold the genuine painting for £169,000.[1]

This case, one of all too many, is an example of the overwhelming plethora of forgeries, fakes, and misattributions that exist throughout the art world. In view of the increasing financial and academic value of an already invaluable collection of masterpieces, it is beyond strange that there has never been a robust standardized protocol for authenticating paintings. Recommendations on the technical requirements for valid written expert opinions for use by the international art community, whether for private purposes or judicial proceedings, remain mere suggestions conveyed to a limited group of researchers and do not in any way represent a consensus that can be generally applied when dealing with issues of authenticity or attribution.

Avoiding such detailed and binding attributional judgments usually results in confining catalogue entries and museum labels to, for example, a simple "Lucas Cranach the Elder," even though, as the present study indicates, most of the works currently attributed to him may very well not have been of his own making. In the absence of a protocol on which there is overwhelming consensus, in an effort to arrive at accurate attributions of paintings, I set out to devise a three-pronged examination through which a work of art can be filtered to achieve that end. My suggested approach includes three complementary elements that together provide an intense and comprehensive examination of the work: art-historical documentation, stylistic connoisseurship,[2] and technical analyses.

The Cranach Workshop

Lucas Cranach was known as the "swiftest of painters," producing some 1,000 extant panel paintings, canvas paintings, drawings, and etchings during his fifty-three-year career (1500–1553), which makes him a remarkable case study. In light of this enormous oeuvre, the obvious question, and the one I address in this article, is: What exactly was Cranach's role in the production of these works, most of which are still widely attributed to him? To answer this question I focus primarily on stylistic comparisons among the works that comprise Cranach's oeuvre in order to assess, analyse, and delineate his style for the ultimate and ambitious objective of determining authorship within Cranach's very busy workshop.

The present study is limited to Cranach's panel paintings. In order to identify and analyze his artistic style, I divided his artistic biography into three periods: (1) his journey to Vienna in 1500–1505, which triggered the explosive character of his vigorous early style and initiated the picturesque School of the Danube, of which eight panel paintings have survived; (2) his call to the Wittenberg castle by the Saxon Elector in 1505. (He remained there he until 1512. Forty-four panel paintings from this period are extant.); (3) his subsequent permanent move to Wittenberg, where he opened his own studio, first in Square Market. (From this period, 132 panel paintings have come down to us. Later, in 1518, on an expanded property that he held until 1550, he produced 673 panel paintings, which are extant. These paintings together number an astonishing 857 panels. I mapped them systematically and studied them using methods of art historical research, connoisseurship, and technical analyses to identify Cranach's personal artistic style. In the pages that follow, I discuss a representative sample from each period. My study indicates that Cranach, one of the richest men in Wittenberg, was interested primarily in accruing wealth, a suggestion that is supported by my conjecture that from 1518 on, he ran a factory-like workshop, supplying apprentices with pigments, designs they could trace, copies, modelling versions, and patterns. In other words, Cranach was an artist turned businessman, worthy of the title "Cranach Inc."

Technical Analyses

The term technical analyses refers to scientific tools such as infrared (IR) and ultraviolet (UV) light, X-ray imagery, and chemical analyses (e.g., of pigments), which enable modern researchers to better understand the processes through which paintings were created and reach a more accurate determination regarding when and where artworks were produced. These tools make a major contribution to ensuring the validity of art authentication and attribution. It should be noted that the analytical diagnostics I used in my research that are not discussed here include X-ray imagery, dendrochronolgy, pigment analyses, UV imaging, and restoration reports. As noted above, the technical analyses I relate to here focus solely on Cranach's panel paintings, specifically the underdrawings, which were usually executed in charcoal or a dark pigment containing carbon, so they can be seen using IR imaging.[3]

Influences on the Way to Wittenberg

Cranach's early style is said to be based on three factors: the influence of his father, Hans, and the contemporary artistic style in Kronach; Dürer's Apocalypse woodcuts; and the impact of the works that were being produced in the region between Regensburg and Vienna around 1500.[4] As there are no extant examples of the "Kronach style" and no works remain from his father's workshop, the first factor remains speculative at best.[5] A search of the region between Regensburg and Vienna for possible influences on Cranach's early style revealed Konrad Witz (1400–1445), Michael Pacher (1435–1498), and Marx Reichlich (1460–1520), but none of these artists had specific links to Cranach, nor is there any visible comparison to be made between them.

A possible influence can be detected in Jan Pollack's *Altarpiece of St. Peter* (Bayerisches National Museum, Munich), where we find the two thieves imaged struggling violently against the cords with which they are tied to the crosses, indicating that this altarpiece, at least, testifies to the unrestrained passion that may have had an effect on the way young Cranach portrayed the two thieves in *The Good Thief on the Cross* and *Thief on the Cross, Facing Left* (Staatliche Museen, Berlin).[6] However, Pollack's possible influence is minimized when we compare Cranach's early style with Dürer's Apocalypse woodcuts.

Dürer's woodcuts show strong contrasts of black and white, loose drawing, and rich contours, all of which appear in Cranach's clearly formulated craftsmanship, which carried over into his paintings, but the latter's depictions are much more extravagant displays, as a comparison between Dürer's and Cranach's renditions of *The Agony in the Garden* (figs. 5.1, 5.2) makes clear. In this work, Cranach surpassed his great Nuremberg contemporary in the expressionistic sweep of his draftsmanship, most clearly seen in the kneeling figure of Christ, in the vigorous plastic accent added by the tree in the foreground, and in the more dramatic contrasts of lights and darks.[7]

This masterful expressionism is evident in all the work Cranach produced in Vienna, with the coloristic brilliance and exuberance that can be seen in *The Wedding Portrait of Dr. Johannes Cuspinian and Anna Cuspinian-Putsch* (Sammlung Oskar Reinhart, Winterthur).[8] *The Lamentation of Christ* (Alte Pinakothek, Munich) is "highly original in its setting and most expressive in the vivid participation of nature . . . [showing] Cranach of the Vienna years . . . as the initiator of the Danube school, which favored . . . romantic landscape motifs . . . and intense coloristic warmth."[9] These expressive characteristics are particularly apparent in comparison with the paintings he later produced in Wittenberg, the birthplace of the Reformation, which reveal fundamental changes in his style.

Although Cranach's paintings viewed in visible light show a consistent style in his Vienna years and suggest that the changes in that style coincided with his arrival in Wittenberg, his underdrawings as revealed through IR imaging indicate the beginning of an evolutionary process in his work. From a connoisseurial point of view, the underdrawings reflect Cranach's spontaneous, individual, unconventional, rhythmic, and inimitable style. Thus, they may be seen as representing the purer form of his artistic process. In the paintings dated between 1500 and 1505, the underdrawings remain stylistically synchronized with the painting itself, whereas after 1505 they differ from the visible painting. Some of the latter IR images still reflect his personal style, indicating that at that point he was still, at the very least, the creator of the underdrawings. From 1518 on we see a distinct transformation in the underdrawings.

Fig. 5.1. Lucas Cranach the Elder, *The Agony in the Garden*, 1502, woodcut. New York, The Metropolitan Museum of Art

Fig. 5.2. Albrecht Dürer, *The Agony in the Garden*, n.d., woodcut. New York, The Metropolitan Museum of Art

Classifying Lucas Cranach the Elder's Work

Attempting to distinguish Cranach's own work as distinct from that of his workshop has been an ongoing concern and became a popular topic of research in the nineteenth and twentieth centuries. Koepplin,[10] Erichsen,[11] and Schade[12] suggest that apprentices were in charge of particular motifs (e.g., heads, landscapes, fabrics), whereas Grimm[13] and Wirth[14] are of the opinion that Cranach and his apprentices also worked on individual paintings. Heydenreich suggests that the division of labor in the workshop took various forms, and notes that "if features of style and quality are uniform and are distinguishable from other paintings, then these may be indicators of the fact that a painting was executed almost single-handedly by the master himself, or by one of his assistants."[15] These features suggest stylistic differences between Cranach and his assistants, but the research concerning attribution in the workshop generally avoids specific attributional judgments. For example, Heydenreich suggests that the headdress of St. Margaret in the *Altarpiece with the Martyrdom of St. Catharine* (Germanisches Nationalmuseum, Nuremberg) was reworked at an advanced stage by a second, superior hand, which leads him to suspect that the reworked part was done by Cranach himself. However, throughout his book, Heydenreich refrains from identifying Cranach's hand based on the style of the painting. How, then, can such a judgment be made? Moreover, looking at other headdresses in the piece, together with the jewelled ensembles, how can an assumption regarding ascription be made about one specific element of the work?

The research concerning attribution in the Cranach workshop based on differentiating artists' hands from one another dates back to Goethe (1749–1832), who attempted to distinguish between Cranach the Elder and his son Lucas. However, this attempt was based on technique rather than style, as he wrote that "there appears to be modelling with paint underneath the glaze, whereas the older pictures are more like drawings glazed with oil. And thus it would not be improbable that these latter paintings could be attributed to Cranach the son, the former to Cranach the father."[16] Cranach's technique was further investigated in 1825 by Schadow (1764–1850), who examined the heads in Cranach's paintings and noted consistency among them, concluding that they were created using mechanical means.[17] In 1851, Schuchardt published two volumes of extensive research regarding commissions, apprentices, and the like in the Cranach workshop, raising the same uncertainties.[18] He tried to resolve the difficulties by determining the painting techniques used in the workshop,[19] but because his hypotheses were fundamentally flawed, his conclusions were misleading. Twenty years later he published a new study regarding the workshop's techniques, this time including primary sources, but his research remains inconclusive. [20] Eisenmann, the first scholar to refer to Cranach's workshop as a production line, wrote in 1877 that Cranach could not possibly have produced all the works attributed to him by himself, and suggested that apprentices created copies based on the master's designs.[21]

After the Cranach exhibition in 1899,[22] the technical aspects of the workshop became the leading topic of associated studies, and stylistic research was summarized as unanswerable.[23] Fleschig addressed the serpent signature that appeared on the paintings and attributed it to Cranach or to his sons, and in so doing ascribed a large number of paintings to his Son's namesake, Hans.[24] Fleschig's hypothesis suffers from one fundamental oversight – different collaborators in the workshop were authorized to sign paintings. Comparison of serpent signatures shows different shapes of the serpent, together with different handwriting for the dates.[25] One can conclude that the signature served as the workshop's trademark and should not be considered an indication of authorship.[26]

The twentieth century saw a heightened interest in the division of labor in the workshop, starting with Friedländer, who remarked: "it is of little importance to know whether the master produced this or that painting entirely or partly with his own hand, or whether one of his workshop assistants had participated in it. The essential thing is the spirit which prevailed in the workshop . . ."[27] Coming from the co-author of Cranach's catalogue raisonné, which was designed mainly to distinguish between originals and copies, this remark is questionable, to say the least. Research regarding theme variations[28] and the influence of patrons on the artistic output of the workshop was also addressed,[29] whereas research regarding technique was rare and focused primarily on dendrochronological investigations,[30] which for the most part were not really viable, so the results proved to be unfounded.

Early twentieth-century attempts to reconstruct the materials and techniques used in Cranach's workshop[31] based on knowledge of the binding media employed proved unsuccessful until the work of White[32] and Campbell.[33] At roughly the same time, taking advantage of new developments in X-radiography and IR reflectography, Benesch was finally able to reveal one of Cranach's underdrawings.[34] This achievement was followed by publications that described the examination of underdrawings and their importance[35] or their insignificance, according to Koepplin and Falk, who maintained that differentiating among the different hands would go against Cranach's work.[36] They insisted that what was significant was the distribution of labour and its importance for understanding workshop procedures and the problems regarding attribution. In contrast, Friedländer and Rosenberg[37] developed the catalogue raisonné by differentiating between originals and copies. However, the catalogue is missing many paintings and marginalizes the significance of knowing who painted what, all of which call into question the validity of the volume itself. The distribution of labour in the Cranach workshop was further examined by Tacke, who analysed the workshop's contract designs in order to determine whether patterns were used.[38] Such research is of great value when studying the IR images from Cranach's oeuvre, which reveal the use of design patterns for tracing.

IR images of the Cranach oeuvre are addressed by Sandner, who underscores their importance in understanding the processes involved in producing a painting.[39] Extensive further research concerning technical aspects of Cranach's work followed Sandner's original study[40] and in 2013 he and Heydenreich published a joint paper that appeared in the *Cranach Digital Archive*.[41]

Nineteenth- and twentieth-century scholars who studied Cranach's workshop apparently concluded that it was impossible to draw conclusions regarding attribution based on stylistic research. Addressing the responsibilities of apprentices and the division of labour in the workshop, coupled with a reluctance to venture concrete judgments, perpetuated the long-standing problem of attribution. Most of the works that came out of the workshop from 1518 on are still credited to Cranach himself, in spite of past catalogue entries that attributed some of them to anonymous masters, his sons, his circle, Hans Abel the Younger, Master of the Mass of St. Gregory, Heinrich Vogtherr, Matthias Grünewald, and many others. It thus seems that the research concerning attributional properties with regard to Cranach is very much subject to debate and remains inconclusive.[42]

Cranach's Style from 1500 to1505

The utterly unrestrained passion evident in the eight extant panel paintings from Cranach's Vienna years that I compared with their IR images shows their expressive characteristics.[43]

All eight paintings were executed with a fluid medium with a brush, or possibly a feather, and are characterized by a free and sketchy air, free-hand underdrawing, thin lines combined with broader lines, and occasional hatching-strokes and washes. As no preparatory sketches survive from this period, based on the IR images it can be assumed that the underdrawings themselves played this role. The IR images show abrupt, expressive lines, not ordered or completely thought through, reinforcing the hypothesis that these are indeed the preliminary sketches.

The contours and volume in all the underdrawings from the Vienna period are indicated with relatively short, curvy lines. Shapes are clarified by repeated lines, which do not necessarily determine the boundaries of the actual painting.[44] These lines are characteristic when addressing specific features of the image, such as the fluidity of fabrics and the execution of the figures. These panel paintings show underdrawings of an almost calligraphic nature, reflecting Cranach's extraordinarily dynamic working method.

Collaborations and Stylistic Changes: 1505–1512

Work from Cranach's period in the Wittenberg castle reflects many changes in his style. Based on the forty-four surviving panels, mostly of religious subjects, we can conclude that this period marks a distinct change in the dramatic tone that infused his earlier works. We can see a combination of flat realism lacking emotional tone and the elongated, slim figures that characterize much of his Wittenberg style, together with an isolation of the figures from the background, much in the manner of a collage. This is evident in the first painting he did after he arrived at the castle, *Altarpiece with the Martyrdom of St. Catharine* (Germanisches Nationalmuseum, Nuremberg), and in such works as the *The Fourteen Helpers in Need* (Kirchgemeinde St. Marien, Torgau), *Portrait of a Man with a Rosary* (the Metropolitan Museum of Art, New York), *Portrait of a Woman* (Kunstmuseum, Basel), among others. Cranach's altered style reflects his ability to adapt to his patrons' desires and inclinations. "Accuracy in depiction, speed of execution, clarity, entertaining variety – such were the virtues most highly esteemed, and Cranach, obedient and reliable servant that he was, was at pains to emphasize those qualities that brought him the greatest reward,"[45] that is, fame and fortune. The question to be asked at this point is whether Cranach's intuitive, spontaneous, and calligraphic style, which was evident in the paintings and the IR images of his Vienna years, changes in the underdrawings as well as the final paintings after 1505.

The answer is yes and no. Of the forty-four panel paintings, twenty display a striking resemblance in the style of their underdrawings to Cranach's Vienna paintings. As it is very unlikely that he developed an entirely different style for some paintings and not for others, it can be inferred with a great degree of certainty that during this period he did not personally execute all of the forty-four paintings attributed to him. Thirteen of these appear to have been done by his apprentices, and eleven seem to be collaborations. Those that can still be attributed to Cranach reflect the style of his early years evident in the IR images, such as the one of *The Fourteen Helpers in Need*. Those that can be attributed to his apprentices can be identified by the distinct tracing lines, which show a simplification of outlines and are limited to essential forms, as seen in *The Judgment of Paris* (Kimbell Art Museum, Fort Worth), *St. Stephen, King of Hungary* (Germanisches Nationalmuseum, Nuremberg), and several others.

It seems that in the works produced in the Wittenberg period, the underdrawings are more definitive than those from his Vienna years, since only slight corrections, if any, are discernible. The idea that the workshop used tracing designs supports the hypothesis that Cranach's expressive style, so beautifully illustrated and highly individual, could not be copied. Thus, his task was not only to produce patterns for tracing, but to simplify and standardize them. This theory gains validity when one analyses his workshop paintings from 1512 onward.

Fizzling Out: 1512–1518

From 1512, the year that Cranach moved into the town of Wittenberg, we can discern a standardized method of production. It appears that Cranach sought to increase the workshop's efficiency by using mostly beech, a series of standard formats, and a modular system for design. As some of these designs have survived, we can see one for religious subjects that was used over and over again. But between 1512 and 1518, a traced composition (discernible by a lack of *pentimenti* and a conformity to the final painting) done in Cranach's distinctive style is still found in the same painting. (I have studied eight such paintings.) This feature points to Cranach's ongoing involvement in the production of such works as *The Virgin and Child with the Infant John the Baptist* (Private Collection), *The Virgin Flanked by Two Female Saints* (Klassik Stiftung Weimar, Museums), and more, alongside his apprentices.

Seven paintings from this period seem to be entirely of his own making, including *The Beheading of St. Catherine* (Archbishop's Castle and Gardens Kromeríz), *The Beheading of St. John the Baptist* (Archbishop's Castle and Gardens Kromeríz), *The Crucifixion* (Statens Museum for Kunst, Copenhagen), *The Mockery of Christ* (Klassik Stiftung Weimar, Museums), and *The Mystic Marriage of Saint Catherine of Alexandria with Saints Dorothy, Margaret, and Barbara* (Szépmuvészeti Múzeum,

Budapest). The remaining one hundred and seventeen panel paintings from this period appear to have been done by his apprentices, as they show traced lines and no *pentimenti*, such as in *The Marriage of St Catherine* (Anhaltische Gemälde-Galerie, Dessau) and *Heinrich the Devout, Duke of Saxony* (Staatliche Kunstsammlungen Dresden). The painting *Pyramus and Thisbe* (Staatsgalerie in der Neuen Residenz Bamberg), which at first glance seems to be in Cranach's style, contains hatching strokes of a type that do not appear in his own works, so it can be attributed to his workshop.

Gone Missing: 1518-1550

Knowing about Cranach's extraordinarily busy and varied career in Wittenberg leads to one cardinal question: Who was responsible for the creation of six hundred and seventy-three panel paintings (that we know of) between 1518 and 1550?

In this vast number of paintings, I have found no stylistic resemblance between Cranach's individual artistic style and the underdrawings that I have examined. Owing to the various types of lines identified in its underdrawings, one painting from this period, *Judith with the Head of Holofernes* (Kunsthistorisches Museum, Vienna), the identity of the artist remains undetermined. For example, Judith's face, nose, eyes, eyelids, and chin are executed with controlled lines, whereas Holofornes's face is sketchier. These differences could point to a possible collaboration, but not with Cranach, as the sketchy lines are themselves somewhat controlled, and do not match Cranach's spontaneous style.

The remaining six hundred and seventy-two panel paintings seem to be workshop productions, mostly of portraits, even though they are still largely attributed to Cranach. These show traced lines in the IR images, no *pentimenti* whatsoever, and highly ordered execution, which suggest pre-existing design patterns and the use of a modular system, which might have made it possible to modify the designs to scale.[46] Evidence of this system strengthens the notion that efficiency was more important than artistic ideals and, together with the IR comparisons, indicates that Cranach was not directly involved in executing the paintings from 1518, and that his workshop was a production line rather than an artistic greenhouse for originality and ingenuity. The use of standard design patterns can be seen in visible light images in, for example, the *Portrait of Leon Braunsberg* (Royal Museums of Fine Arts of Belgium, Brussels), *Portrait of Luther as Junker Jörg* (Museum der bildenden Künste, Leipzig), and others, which reflect a prototypical use of portrait design.

As Cranach used a modular system for his designs, it may be assumed that he did the same when it came to painting on the underdrawings, working on each section separately. But since the palette itself could only hold a limited quantity of paint, it is not entirely unreasonable to imagine that different sections of the same painting were painted by different artists, each using his own palette on the section for which he was responsible.[47] As Cranach's workshop grew, production increased, and the above-mentioned techniques helped this enterprise maintain its volume of sales, and some might say its quality as well. Since minor corrections can be discerned in the paintings produced from 1518 to1550, it may be assumed that the workshop, with its factory-like conditions, supplied Cranach's apprentices with pigments and designs that they could trace, making Cranach an artist turned businessman, justifying the title "Cranach Inc."

Conclusion

Cranach's move to his private studio; his use of modular designs, standardized panels, and ready-made pigments; his purchase of large quantities of pigments from the pharmacy at low cost, since several painting materials were considered pharmaceutical items in the sixteenth century; and a task force of at least thirty at any given time are all indications that Cranach's workshop was a production line. That being the case, modern scholarship should engage in studies to re-examine, relabel, and rediscover one of the greatest German painters of the Renaissance.

Notes

[1] Will Bennett, Two versions of Gauguin work on sale at same time, *The Telegraph*, March 12, 2004. http://www.telegraph.co.uk/news/worldnews/northamerica/usa/1456587/Two-versions-of-Gauguin-work-on-sale-at-same-time.html. Accessed June 10, 2015.

[2] According to Carol Gibson-Wood, the practice of connoisseurship is the "identification of authorship by examining a work's style." See her "Studies in the Theory of Connoisseurship from Vasari to Morelli" (PhD diss., The Warburg Institute of London, 1982): 6. Although this twentieth-century definition encapsulates the essence of what connoisseurship *is*, it cannot be fully understood without addressing its evolution and its contribution to the development of art historical research. My research, which owing to the space limitations of this chapter, cannot be fully addressed here, traces the practice of connoisseurship from Giorgio Vasari to Max Friedländer.

[3] "The National Gallery Glossary:"

http://www.nationalgallery.org.uk/paintings/glossary/infrared/*/chooseLetter/I. Accessed August 7, 2016.

[4] W. L. M. Burke, "Lucas Cranach the Elder," *The Art Bulletin* 18, no. 1 (1936): 27.

[5] Ibid., 28. Burke suggests that the Nuremberg style might have influenced Lucas Cranach the Elder through the works of Hans Pleydenwurff (1420–1472) and Graf Löwenstein, among others, but there is no degree of certainty regarding their effect, if any, on Cranach.

[6] Curt Glazer, *Lukas Cranach* (Leipzig: Im Insel-Verlag, 1923), 18.

[7] Jakob Rosenberg, "Lucas Cranach the Elder: A Critical Appreciation," *Record of the Art Museum, Princeton University*28, no. 1 (1969): 33.

[8] On the iconographic symbolism and the humanistic circle of Vienna displayed in these portraits, see Dieter Koepplin, *Cranachs Ehebildnis des Johannes Cuspinian von 1502: Seine christlich-humanistische Bedeutung* (Düsseldorf: Stehle, 1973).

[9] Jakob Rosenberg, "Lucas Cranach the Elder": 36. Because of its expressive intensity, this painting has been falsely attributed twice: first to Matthias Grünewald in 1855 by Adolph Bayersdorfer in *Verzeichniss der in der königlichen Gallerie zu Schleissheim aufgestellten Gemälde*, cat. no. 184 (Munich: Knorr and Hirth, 1885) and then to Wolf Huber in 1892 by Wilhelm Schmidt in "Wolf Huber und M. Grünewal," *Zeitschrift für bildende Kunst. Kunstchronik* 3, no. 27 (1892): 118. It was only in 1895 that it was attributed to Cranach by Franz Rieffel in "Kleine kunstwissenschaftliche Controversfragen," *Repertorium für Kunstwissenschaft* 18 (1895): 425.

[10] D Dieter Koepplin and Tilman Falk, *Lukas Cranach: Gemälde, Zeichnungen, Druckgraphik: Ausstellung im Kunstmuseum Basel, 15. Juni bis 8. September 1974*, exhibition catalogue (Basel: Birkhäuser, 1974–1976): 37.

[11] Johannes Erichsen, "Altäre Lucas Cranachs und seiner Werkstatt vor der Reformation," in *Lucas Cranach: Ein Maler-Unternehmer aus Franken*. Eds. C. Grimm, J. Erichsen, and E. Brockhoff. Augsburg: Haus der Bayerischen Geschichte; Regensburg: F. Pustet, 1994). Exhibition Catalogue, 150–165.

[12] Werner Schade, "Beobachtungen zur Arbeitsweise in Cranachs Werstatt," *Kunsterziehung* 7, no. 8 (1972): 7–10.

[13] Claus Grimm, "Lucas Cranach 1994," in *Lucas Cranach: Ein Maler-Unternehmer aus Franken,* eds. C. Grimm, J. Erichsen and E. Brockhoff (Augsburg: Haus der Bayerischen Geschichte; Regensburg: F. Pustet, 1994). Exhibition Catalogue, 18–43.

[14] Jean Wirth, "Cranach Reconsidéré," *Revue de l'Art* 37 (1977): 83–99.

[15] Gunnar Heydenreich, *Lucas Cranach the Elder: Painting Materials, Techniques and Workshop Practice* (Amsterdam: Amsterdam University Press, 2007), 290.

[16] Johann Wolfgang Goethe, March 22, 1815, *Morgenblatt für gebildete Stände*, no. 69, 273–274, quoted in Susanne Heiland, ""Wahrheit und Legende: Die Geschichte des Bilderfundes" and "Archivalien und Texte zum Bilderfund von 1815"," in *Vergessene Altdeutsche Gemälde,* ed. Herwig Guratzsch, (Museum der bildenden Künste Leipzig and Heidelberg,1997). Exh. Cat., 155.

[17] Johann Gottfried Schadow, *Wittenbergs Denkmäler der Bildnerei, Baukunst und Malerei, mit historischen und artistischen Erläuterungen* (Wittenberg: Zimmermannische Buchhandlung, 1825), 96.
[18] Christian Schuchardt, *Lucas Cranach des Alteren Leben und Werke*, parts I, II (Leipzig: F. A. Brockhaus, 1851).
[19] Ibid., I, 89.
[20] Christian Schuchardt, *Lucas Cranach des Alteren Leben und Werke* (Leipzig: F. A. Brockhaus, 1871).
[21] Oskar Eisenmann, "Lucas Cranach," in *Kunst und Künstler*, ed. R. Dohme (Leipzig: E. A. Seemann, 1877), 34.
[22] Karl Woermann, *Deutsche Kunst Ausstellung Dresden, 1899: Abteilung Cranach Ausstellung Wissenschaftliches Verzeichnis Der Ausgestellten Werke* (Dresden-Blasewitz: A. Arnold, 1899).
[23] Karl Woermann, "Die Dresdner Cranach-Austellung," *Zeitschrift für bildende Kunst*, NF 11 (1900): 88; Eduard Fleschig, *Tafelbilder Lucas Cranachs d.Ä. und seiner Werkstatt* (Leipzig: E. A. Seemann, 1900): 7. Woermann stated that only the best works should be attributed to Cranach, and Fleschig maintained that he could not recognize Cranach's works after 1520.
[24] Fleschig. *Tafelbilder Lucas Cranachs*, 7–9.
[25] Heydenreich, 293.
[26] For more on the personal symbol and its emergence in the North, see Joseph Leo Koerner, *The Moment of Self-Portraiture in German Renaissance Art* (Chicago: University of Chicago Press, 1993), 203–223.
[27] Max J. Friedländer, "Die frühesten Werke Cranachs," *Jahrbuch der Preußischen Kunstsammlungen* 23 (1902): 228.
[28] Ibid., 2.
[29] Ernst Heidrich, *Die Altdeutsche Malerei* (Jena: Diederichs, 1909), 8.
[30] Fleschig, *Tafelbilder Lucas Cranachs*, 9–33.
[31] Max Doerner, *Malmaterial und seine Verwendung im Bilde* (Stuttgart: Enke, 1976), 187–191.
[32] Raymond White and Jennifer Pilc, "Analysis of Paint Media," *National Gallery Technical Bulletin* 16 (1995): 88–89.
[33] Lorne Campbell, Susan Foister, and Ashok Roy, "Methods and Materials of Northern European Painting in the National Gallery, 1400–1500," *National Gallery Technical Bulletin*. 18 (1997): 6–55.
[34] Otto Benesch, "Zur Altösterreichischen Tafelmalerei. Die Anfänge Lukas Cranachs," *Jahrbuch der Kunsthistorischen Sammlungen Wien* 38, no. 2 (1928): 78, 81.
[35] Werner Schade, *Die Malerfamilie Cranach* (Dresden: Verlag der Kunst, 1974); Schade 7–10; Werner Schade, "Beobachtungen an Vorzeichnungen von Cranach-Bildern," in *Akten des Kolloquiums zur Basler Cranach-Ausstellung 1974*, ed. D. Koepplin (Basel: Kunstmuseum, 1977). Exhibition catalogue ,11–12; J. R. J. Asperen de Boer, "The Development of Infrared Techniques in the Examination of Paintings," in *Unsichtbare Meisterzeichnungen auf dem Malgrund: Cranach und*

seine Zeitgenossen, eds. I. Sandner, W. S. Eisenach, and F. Köln (Regensburg: Schnell and Steiner, 1998). Exh. Cat., 21–25.

[36] Koepplin and Falk, 12.

[37] Max J. Friedländer and Jakob Rosenberg, *The Paintings of Lucas Cranach* (London: Philip Wilson Publishers Ltd., for Sotheby Parke Bernet, 1978).

[38] Andreas Tacke, *Cranach: Meisterwerke auf Vorrat: Die Erlanger Handzeichnungen der Universitätsbibliothek. Bestands- und Ausstellungskatalog* (Munich: Form-Druck, 1994), 51–66.

[39] Ingo Sandner, "Die Analyse von Unterzeichnungen auf Gemälden mit Hilfe der Infrarotreflektografie am Beispiel von Werken Lucas Cranach d. Ä.," in *Deutsche Gesellschaft für Zerstörungsfreie Prüfung*, e.V. 4th International Conference on Non-Destructive Testing of Works of Arts (Berlin 1994): 193–202; Ingo Sandner, "Cranach als Zeichner auf dem Makgrund," in *Unsichtbare Meisterzeichnungen auf dem Malgrund. Cranach und seine Zeitgenossen*, eds. Ingo Sandner, Wartburg-Stiftung Eisenach, and Fachhochschule Köln (Regensburg: Schnell and Steiner, 1998). Exh. Cat., 83–95.

[40] Iris Ritschel, "Das Gemälde 'Die Vierzehn Nothelfer' und 'Christus als Schmerzensmann' in der Marienkirche zu Torgau," *Denkmalpflege* (1995): 38–62; Iris Ritschel, "Der Frankfurter "Annenaltar" von Lucas Cranach dem Älteren aus dem Jahre 1509: Ein Werk aus der Marienkirche zu Torgau?," *Kleine Schriften des Torgauer Geschichtsvereins* 6 (1996): 7–26; Gunnar Heydenreich, "Das Bildnis einer jungen Dame: Eine Vorlage für die Darstellung der hl. Margareta auf dem Dresdner Katharinenaltar (1506) von Lucas Cranach d.Ä.?" in *Monatsanzeiger. Museen und Ausstellungen in Nürnberg* 195, no. 6 (1997), special issue: Germanisches Nationalmuseum, ed. Georg Ulrich Großmann, : 7–11; Gunnar Heydenreich, "Herstellung, Grundierung und Rahmung der Holzbildträger in den Werkstätten Lucas Cranachs d.Ä.," in *Unsichtbare Meisterzeichnungen auf dem Malgrund: Cranach und seine Zeitgenossen*, eds. I. Sandner, W. S. Eisenach, and F. S. Köln (Regensburg: Schnell and Steiner, 1998). Exhibition Catalogue, 181–200; Gunnar Heydenreich, "Artistic Exchange & Experimental Variation: Studies in the Workshop Practice of Lucas Cranach the Elder," in *Painting Techniques, History, Materials and Studio Practice*, eds. Ashok Roy and Perry Smith, (London: International Institute for Conservation of Historic and Artistic Works, 1998), 106–111.

[41] Ingo Sandner and Gunnar Heydenreich, "Cranach als Zeichner auf dem Malgrund–Auswertung der Untersuchungen im infraroten Strahlenbereich," *Cranach Digital Archive*, 2013. The Cranach Digital Archive (CDA) is an interdisciplinary collaborative research resource that provides access to art historical, technical, and conservation information on paintings by Lucas Cranach (c.1472–1553), his sons, and his workshop. The repository presently provides data on more than 1,000 paintings including some 10,000 images and documents from 127 contributing institutions. See http://www.lucascranach.org/

[42] I believe that aspects of litigation prevent art historians from unequivocally attributing artworks.

[43] All the paintings and their respective IR images can be seen in the CDA. See note 41.

[44] Sandner and Heydenreich, "Cranach als Zeichner auf dem Malgrund," 7–8.

[45] Friedländer and Rosenberg, *The Paintings of Lucas Cranach*, 18.

[46] David Bomford, ed., *Underdrawings in Renaissance Paintings* (London: National Gallery Co. and New Haven, CT., Yale University Press, 2002), 22.

[47] Jo Kirby, "The Painter's Trade in the Seventeenth Century: Theory and Practice," *National Gallery Technical Bulletin* 20 (1999): 15.

PART 3:

CREATION AND EMBODIMENT

CHAPTER SIX

JEWISH WOMEN, JEWISH LIGHT

SUSAN NASHMAN FRAIMAN

Judaism is known for several unique lighting implements:[1] the seven-branched Temple Menorah,[2] which stood in the tabernacle and later in the Temple; the eternal light in the synagogue; the Hanukkah lamp;[3] the Sabbath lamp, lit on Friday evenings before sundown to usher in the Sabbath; as well as special candles, such as the Havdalah candle[4] and those lit on Yom Kippur. Women are associated in particular with the lighting of the Sabbath lamp, as mentioned first in the Mishna in Tractate Shabbat 2:6. Today, the most common way to inaugurate the Sabbath is with the lighting of candles made either of paraffin or stearin, but until the mid-twentieth century, hanging oil lamps were in use in many European communities. Until the modern era, women had little or no choice regarding the kind of lighting implements they used to usher in the Sabbath. This article will discuss the origins of the Sabbath lamp, as well as extra-halachic ways Jewish women found to add meaning to the obligation to light the Sabbath lamp, while both adhering to Halachic norms and instigating cultural innovation.

Surprisingly, unlike most of the Jewish commandments, or *mitzvoth*, the commandment to kindle the Sabbath lights does not appear explicitly in the first five books of the Hebrew Bible, known in Hebrew as the Torah, as opposed to other commandments, like, for instance, "Honor thy father and mother." Rather, it is derived from the injunction in Exodus 35: 3: "You shall kindle no fire throughout your settlements on the Sabbath day."[5] From this prohibition, the rabbis understood that light, heat, and warm food needed to be provided before the onset of the Sabbath.[6]

Women's connection to the obligation to light the Sabbath lamps first appears in a rabbinic text in the second century CE Mishna, yet the practice of Sabbath lighting was recorded by early Roman writers and by Josephus already in the first century of the Common Era. The actual text of the Mishna associating women with the Sabbath light prescribes a harsh punishment for the lack of observance—death in childbirth: "On account of three transgressions do women die in childbirth: because they are not meticulous in the laws of menstrual separation, in the dough offering [*hallah*], and in the kindling of the lamp [for the Sabbath]."[7]

The extreme nature of thc punishment for the failure to light the Sabbath candles may have been a way to equate the commandment to light with the other two commandments mentioned in the same Mishna, which are both biblically ordained and therefore have a slightly different status.[8] It may also have been a way to ensure that there was light in the home on the Sabbath, because the rabbis endeavoured to make Sabbath observance pleasurable and not experienced as a punishment, according to the principle of "You shall live by them," in Leviticus 18:5: "Ye shall keep My laws, and My rules, by the pursuit of which man shall live; I am the LORD."

For this reason there is an elaborate set of laws that enable Jews to have hot food on the Sabbath, as well as light. Much rabbinical writing is devoted to this subject, since if candle lighting is not specifically commanded, then what are the reasons for it? Rabbinic writings proffer at least three: pleasure, domestic peace, and honouring the Sabbath. With no light in the home, as mentioned above, the Sabbath experience, which was supposed to be pleasurable, would be significantly less so. Lack of light could also lead to accidents and resulting domestic discord. The third reason, honouring the Sabbath, has actually become the main rationale for Sabbath candle lighting today, where electric lighting is the norm, and the difficulty and expense of lighting the home, which prevailed for close to a thousand years, have completely disappeared.

Thus, the provision of light for domestic harmony became the prerogative of the woman, who is seen in rabbinic sources as responsible for the home.[9] Nonetheless, if no woman is present, men are obligated to light the Sabbath lamp as an integral part of Sabbath observance.[10]

What kind of lighting implement was used? The development of the European lighting implement for the Sabbath is a fascinating one, but one in which women played little role. As opposed to the Hanukkah menorah, there were only a few requirements for the implement used to light for the Sabbath. First, olive oil was the preferred material for lighting, since it was used in the Temple and burnt the best, but certain other oils were also permissible, as long as they did not produce a bad odor (Mishna Shabbat 2:2). In addition, there had to be at least two wicks,[11] and the lamp had to be especially designated for the Sabbath.[12] The lack of specificity with regard to the design of the Sabbath lamp may point to a rabbinic desire to make the weekly observance as easy as possible, and leaves open myriad possibilities. Indeed, research shows that the development of the Sabbath lamp is situated at the intersection of Jewish law and material culture. Local climate and custom certainly affected the development of these lamps. The availability of different lighting materials in different areas of Europe and, by extension, other areas inhabited by the Jews, also played a role in the form of their lamps. To discuss the development of the European lamps briefly, there is evidence that hanging glass lamps containing a layer of oil atop water were the most common form of lamp used at the beginning of the Middle Ages in both Europe and the East, and that this changed towards the end of the thirteenth century in Europe.

One of the interesting gaps between halachic prescription and the evidence from research on material culture is the discrepancy between the number of lamps/candles or wicks required to be lit and the actual number that were lit. There are two issues here. One is that the Hebrew word for candle, *ner,* means both lamp and wick, so when the thirteenth century rabbi Eliezer ben Yoel Ha-Levi (1140–1225), known by the acronym as the Ravyah, says that two *nerot* are the halachic minimum, it is difficult to determine exactly what he meant.[13] The other issue is that the evidence of material culture contradicts his dictum; multi-wicked spouted lamps that contained more than two wicks or candles were used. Other Halachic sources as well as material evidence from that period reflect the move toward a multi-wicked lamp that was used specifically for Sabbath lighting. The earliest example of such a Jewish lamp was found in an excavation near Cologne, Germany and is now in the collection of the Jewish Museum of New York City. Most likely this form is based on a lamp used by French nobility called the *croiseul*, which itself was based on more elaborate ecclesiastical lamps. This form of lamp fell out of general domestic use, but was adopted by the Jews for Sabbath lighting. This lamp, in fact, eventually became an attribute of the Jews, and was known to non-Jews as a *Judenstern*. The most common number of branches was eight,[14] but there were also six- and ten-branched lamps, usually made of brass (fig. 6.1 Woman Lighting the Sabbath Lamp, *Book of Customs,* Amsterdam, woodcut, (Amsterdam: Isaac de Cordova, 1723); Courtesy of the University of Chicago Library, Harry and Branka Sondheim Jewish Heritage Collection.)

A few prominent Jewish families had ornate lamps made of silver.[15] In Jewish communities influenced by the *Kabbalah* (Jewish mysticism), such as the Portuguese Jewish community of Amsterdam, there were seven-wicked lamps with less clearly defined arms, reflecting the mystical significance of the number seven.[16]

Fig. 6.1 Woman Lighting the Sabbath Lamp, *Book of Customs,* woodcut by anonymous artist from the *Book of Customs,* (Amsterdam: Isaac de Cordova), 1723.

To return to the connection of women to the Sabbath lighting, if women did not choose the implements they used, were there ways in which they were active participants in their performance of this commandment? Sources reveal that women taught other women about the laws of lighting the lamps and the blessing to be recited.[17] How did these women perceive their actions and role?

It is useful here to compare Sabbath candle-lighting with the lighting of the Yom Kippur candles. The custom of bringing long candles to the synagogue on the eve of Yom Kippur has never been systematically researched. Rabbinic texts mention this custom already in the Talmud.[19]

While there are many sources that testify to this custom, there seem to have been two traditions—bringing to synagogue one candle that would burn for the entire day and bringing a candle that would burn for a longer period of time, either from Rosh HaShanah until Yom Kippur, or from Yom Kippur until Hoshanah Rabbah.[20] The former custom, which seems to have been more common in eastern Europe, involved lighting two candles, one at home and one in the synagogue. This may derive from the fact that in some communities the Yom Kippur candle was brought home at the end of the fast to make Havdalah.[21] However, one can imagine that there were technical difficulties with keeping the flame lit between the synagogue and the home, and thus other sources suggest that there also be a candle at home for lighting.[22] The Jewish apostate, Antonius Margaritha (1490–1542) describes in *Das Ganz Judische Glaube* (1531) the custom of the Yom Kippur candle: on the day before Yom Kippur, the Jews would come to the synagogue and place their candles, which had to burn for at least an entire day, along the wall of the synagogue. However, he also describes another custom: "I saw men who would bring candles that weighed 24, 28 or even 30 pounds."[23]

The depiction of the candles that appear in Margarita's work (fig. 6.2) based on an earlier depiction in the anti-Jewish work *Die Judenbeicht* (1508) by another Jewish apostate, Johannes Pfefferkorn (1469–1523), is more likely that of the candles that burned for 25 hours in the synagogue. The persistence of this custom is attested to by depictions of these candles in both nineteenth- and twentieth-century works. The painter Moritz Oppenheim included the candles in his depiction of the Rothschild family at prayer on Yom Kippur,[24] and in his nostalgic depictions of German-Jewish life, *Bilder aus dem altjüdischen Familienleben* (fig. 6.3).[25] The two twentieth-century depictions, both from 1906, are by Leopold Pilichowski, and Maurycy Minkowski.[26] In the Oppenheim depiction of the synagogue, one clearly sees a ledge where the candles were placed, as in the sixteenth-century print. One can also see such a ledge in pre-war prints and photographs of the Worms Synagogue.[27]

Interestingly enough, only men were obligated to take part in this Yom Kippur observance. One of the explanations given is that the numerical value of the Hebrew word for candle, *ner*, equals the number of limbs in a man's body, plus two: his spirit and soul (*ruah v'neshamah*).[28] Another explanation is based on a verse from Proverbs: "For the commandment is a lamp, the teaching is a light." (Proverbs 6:23). The light is equated with the Torah, which women were exempted from studying. However, women were involved in making these candles, as testimony has shown. The practice of lighting two candles seems to have been most pronounced in eastern Europe, where candle making fell to the mother of the household on the eve of Yom Kippur. These candles were known as the "memorial candle" and the "candle for the living." Both were large and needed to burn for the entire day of the fast. The memorial candle was made in memory of family members who had died, and the custom of lighting such a candle persists to this day. The other candle, the "candle for the living" (or for the "healthy"), was also made by women and consisted of long candle that would burn throughout Yom Kippur, but would be placed in the synagogue, not kept at home. This custom may still be practiced in some communities.[29]

Fig. 6.2 Jews Praying in the Synagogue on Yom Kippur, woodcut by anonymous artist, from Anton Margarita, *Der gantz Jüdisch glaub* (Augsburg: Heynrich Steyner), 1531.

Fig. 6.3 Moritz Oppenheim (1800-1882), "The Eve of the Day of Atonement," lithograph, *Bilder aus dem altjüdischen Familienleben* (Frankfurt: Keller, 1886), unnumbered page.

The symbolism of lighting a candle for the soul in Judaism is based on several texts, for instance: (Proverbs 20:27) "The life breath of man is the lamp of the Lord, revealing all his inmost parts."[30] Since the "candle for the living" was lit in the synagogue, if the candle went out during Yom

Kippur, it was considered a bad omen. The rabbis went so far as to suggest that if it did go out during Yom Kippur, it should be relit after the holiday was over.[31] The worry associated with these candles was so great that synagogue beadles were encouraged to mix up the candles so that the person who brought the candle would not know which one was his.[32]

Several memoirs by have described the making of the Yom Kippur candles in Eastern Europe. [33] The earliest and perhaps most well-known account by a woman is the memoir of Pauline Wengeroff (1833–1916), *Memoirs of a Grandmother,* in which she describes the practice.[34] The candles were about a meter long, and while reciting wishes for her living family, the mother of the household, aided by another woman (known as the *gabete*, a woman with a ritual role) would pull the cord used to make the wicks. Each thread in the wick symbolized another member of the family, and while the cord was pulled, special wishes for that member of the family would be recited, along with special supplications.[35] The memorial candle was made in much the same way, with the exception that the supplications would be directed to the matriarchs and patriarchs and deceased family members. Some accounts of this custom mention that the circumference of the graves of the ancestors prayed for were measured with string that was used to make the wicks.[36] The measuring of the circumference was likely an adaptation of a Catholic practice that perhaps originated in France. The Christian custom was performed when a community was in distress, for example during an attack of the plague.[37] A very long wick, usually measured by the circumference of the city, was made for a taper, which would burn for months in the church in the wake of misfortune or in order to prevent disasters. Another Catholic custom involved taking the measure of the height of a person seeking mercy with a wick, and making a candle with it on their behalf.[38] These large candles may have influenced the custom of measuring the graves of the ancestors, as mentioned above, and are but one example of the synergy between rabbinic tradition and the surrounding culture.[39]

The making of the Yom Kippur candles by women epitomizes their role of women as facilitators for their husbands and sons. While women could not study Torah, they could gain merit through her husband and sons. A woman who sent her sons and husband to study Torah gained merit vicariously, as mentioned in the Talmud.[40] According to Jewish law, women are exempted from many of the commandments, particularly the positive time-bound commandments, yet paradoxically the lighting of the Sabbath lamp is a time-bound commandment that women are obligated to perform.[41] While the making of the Yom Kippur candle grants vicarious virtue to women, in lighting the Sabbath lamp, the woman is a direct actor in ushering in the Sabbath. It is interesting to note that the connection to women is not specified in the mishnah that links women to candle-lighting. The reasons only appear in later *midrashim*,[42] which were in turn cited in the Zohar.[43] One of the reasons given for the association of women with candle-lighting has to do with Eve's actions in the Garden of Eden. Eve, by giving the apple to Adam, brought death into the world, and in so doing "extinguished the light of the world." Here the light refers both to Adam and to light as a metaphor for the spiritual and the good. Therefore, women's requirement to light Sabbath candles is seen as a *tikkun*, a reparative act for this original error. The notion of lighting the lamp as punishment has been discussed by Judith Baskin.[44] However, as opposed to rabbinic ambivalence regarding women, there is no ambivalence regarding light and the making of light. Light always has a positive connotation; for instance, light is associated with the soul, as we saw above. Folk belief is also an important factor here—for instance, in German-Jewish folklore, light was believed to expel demons who would otherwise destroy the delight of Friday night.[45] Hence the lighting of the Sabbath lamp should not solely be interpreted as a punishment, but also as a *tikkun*, a repair of something that is broken in the world. The other side of the coin, of course, is that women are responsible for the home and therefore candle lighting, along with other domestic chores, is in their purview. But in candle-lighting, women are also in the role of active participants; their lighting of the lamp is essential for everyone's (including their own) enjoyment of the Sabbath.

Interestingly, women, did not, by and large, focus on the negative reasons for candle lighting. As in the making of the Yom Kippur candle, women recited special prayers before and/or after candle lighting. These prayers were extra-halachic, meaning that they were not required, nor did they have a fixed text. There are several known variants of these prayers, which are called *tehinnot* or supplications, and Chava Weissler has written about them extensively.[46] The custom of reciting these supplications actually originated in Spain and migrated north.[47] The tradition of women writing prayers in general also seems to have originated in Spain.[48] Such prayers were written not only to be said at the time of candle-lighting, but also to fill a variety of needs not covered by the traditional prayers. There were also supplications written for men. The earliest book of such supplications was published in Amsterdam,[49] another possible indication of Iberian influence, considering the strong Portuguese community there. Their writers were mostly anonymous, and probably predominantly men, at least until the end of the eighteenth century, when we have evidence of women writers.[50] An examination of the prayers written by women shows that almost none mentions that women need to make *tikkun* for bringing death into the world. Instead, the women writers of these supplicatory prayers see themselves as acting much in the same way as the High Priest in the Temple. One such text, written by Sarah Bat Tovim in the first half of the eighteenth century asks: "May you G-d, receive our light the way you received the light of the *High Priest in the Temple*."[51] A painting in the collection of the Prague Jewish Museum shows a woman actually lighting a seven-branched lamp.[52] The case for the importance and efficacy of women's prayers was made by another eighteenth-century woman, Sarah Rebecca Rachel Leah Horowitz, who composed a special prayer to be recited by women in the synagogue on the Sabbath of the blessing of the New Moon. She demonstrated a breadth of learning in her introduction, explaining why women's prayer was essential.[53]

An opportunity for cultural reframing came with the nineteenth-century revolution in domestic lighting, which led to the change in the function of Sabbath candle-lighting. The Sabbath candles were no longer needed to actually light the home; indeed, there were even halachic discussions as to whether one could recite the blessing on electric lights.[54] This technological development led to a change in the performance of candle-lighting. The fact that most modern candlesticks designed for a Jewish audience came in pairs was based on the early halachic dictum by Rabbi Eliezer b. Joel HaLevi mentioned above that stated that two lights or wicks were required, one for the first reference in the Torah of the commandment to keep the Sabbath and one for the second.[55] So, while two flames were the halachic minimum, in fact, the most commonly used vessels had places for more than two (see above). Paulina Wengeroff describes how she, her mother, and her sisters would light the multi-branched Sabbath lamp; each one would light candles until it was completely lit. We see the continuation of this custom in a painting by the Swedish-Jewish painter Geskel Saloman (fig. 6.4).

The need to create light on the Sabbath as one of the main functions of candle-lighting was largely obviated with the advent of other forms of home lighting, especially electricity, but also gas and kerosene. While there is halachic agreement that electric lights may be blessed as Sabbath lights,[56] in fact, the desire to light candles has not diminished with time and candle-lighting has evolved from serving the practical function of illuminating the home to fulfilling a symbolic purpose. The custom of lighting multiple Sabbath lights in the home has also evolved into a symbol, with additional lights being added for the birth of each child. Women added a new meaning to lighting; this is a custom that was promulgated by women and is practiced by many today.[57]

Fig. 6.4 Geskel Saloman (1821-1902), *The Blessing of the Sabbath Candles*, 1898, oil on canvas, 120 X 140 cm.; courtesy of the Jewish Community of Stockholm.

I have attempted to show the complexity of women's role in Judaism in connection with two ceremonies, lighting the Sabbath lamp and making the Yom Kippur candles. In both cases, women play or played a pivotal role as both facilitators and participants. In both cases, I would argue, women found their own meaning, independent of rabbinical interpretation, which in some cases involved denigration. This kind of grass-roots practice had halachic repercussions, and often, custom became the basis for redrawing the lines of what was permissible or even normative.[58] The ceremonies of creating light in the Jewish home and synagogue were in fact given to women to perform and women created meaning within them, even when they did not create the form. Both the composition of prayer, and the addition of new meaning to candle-lighting were part of the process of the modernization of the Jewish community, as historian Murray Jay (Moshe) Rosman has observed: "Although not usually noted by conventional portrayals, the Jewish modern period was characterized by the continual, if very slow, transformation of Jewish women's role from that of cultural bystanders to that of participants, and from social facilitators of the activities of their menfolk to performers who engage in the most socially valued activities."[59] The attribution of new meanings to candle-lighting reveal both Jewish women's attachment to this commandment and to their ritual creativity in its performance.

Notes

[1] This paper is based on my doctoral thesis, Susan Nashman-Fraiman, "The Sabbath Lamp: Development of the Implements and Customs for Lighting the Sabbath Lights among the Jews of Ashkenaz" (Jerusalem: Hebrew University, 2013) (Hebrew).

[2] See Steven Fine, *The Menorah: From the Bible to Modern Israel* (Cambridge: Harvard Press, 2016).

[3] A lamp that has eight or nine branches and is lit to commemorate the rededication of the Temple by the Hasmoneans in the second century BCE. See *Encyclopaedia Judaica*, s.v. Hanukkah Lamp (Jerusalem: Keter, 1972), vol. 7, cols. 1287–1315.

[4] Michele Klein, "The Candle of Distinction: a Cultural Biography of the Havdalah Light," *Images* 8 (2014): 3–24.

[5] *Tanakh: A New Translation of the Holy Scriptures* (Philadelphia: Jewish Publication Society, 1985), 141. All Biblical translations from Hebrew will be taken from this edition.

[6] One branch of Judaism that did not agree with this interpretation was Karaism. Karaites do not accept the oral tradition of the Talmud and some of them sit in darkness on the Sabbath to this day. See *Encyclopaedia Judaica* s.v. Karaites (Jerusalem: Keter, 1972), vol. 10, col. 779.

[7] Based on Jacob Neusner, trans., *The Mishnah: A New Translation* (New Haven: Yale University Press, 1988), 181.

[8] For a discussion of the different statuses, see Menachem Elon, Bernard Auerbuch, and Melvin J. Sykes, trans., *Jewish Law History, Sources, Principles* (Philadelphia and Jerusalem: The Jewish Publication Society, 1994), 207–224.

[9] Rambam: *Mishneh Torah Shabbat* 5:3; *Shulḥan Arukh, Oraḥ Ḥayyim* 263:3.

[10] Yehoshua Neuwirth, *Shmirath Shabbath A Guide to the Practical Observance of the Shabbath* (Jerusalem: Feldheim, 1997) Vol. 3, 731-732.

[11] Nashman Fraiman, "The Sabbath Lamp," 121.

[12] Ibid., 126.

[13] Avigdor Aptowitzer, ed., *Sefer HaRavyah*, Part 1 (Jerusalem: Harry Fischel Institute, 1964): Responsa 199 (Hebrew). For a more detailed discussion of this point see Nashman-Fraiman, "The Sabbath Lamp," 29.

[14] Ibid., 197–202.

[15] Ibid., 209–228; For one example, See Vivian Mann, ed., *Treasures of the Jewish Museum* (New York: Universe Press, 1986), 80–81.

[16] Nashman Fraiman, "The Sabbath Lamp," 32.

[17] Jacob Tam (1100-1171), *Sefer HaYashar,* ed. Shraga Rosenthal (Zvi Hirsch: Berlin, 1898), responsa 48: 6, 99 (Hebrew).

[19] *Babylonian Talmud, Horayot*, 12a; Keritot, 5b; See Daniel Sperber, *The Jewish Life Cycle* (Ramat Gan: Bar-Ilan University Press, 2008), 569–570. On at least one thirteenth-century rabbi who discussed this, see *HaRavyah*, *Rosh Hashana*, responsa 547 (Hebrew).

[20] Joseph Bar Moses *Leket Yosher Customs Ritual Decisions and Responsa of Rabbi Israel Isserlein*, ed. Jacob Freimann, (New York: Menorah, 1959), 142 (Hebrew).

[21] Ibid.

[22] Perhaps this is the origin of the development of the use of two candles, one lit at home and one lit in the synagogue. Shlomo Joseph Zevin, ed. *Talmudic Encyclopedia* (Jerusalem: Talmudic Encyclopedia Institute, 1984), Vol. 4, 554 (Hebrew).

[23] These were apparently the large candles that burned until the last night of the Sukkot holiday, *Hoshana Rabbah*, as mentioned by a fifteenth century source, the *Leket Yosher* (see n. 20). Quoted in Richard Krautheimer *Mittelalterliche Synagogen* (Berlin: Frankfurter Verlags-Anstalt, 1927), 116–117. (Link to images from the Princeton University Graphic Arts Collection: https://graphicarts.princeton.edu/2013/11/19/der-gantz-judisch-glaub/. Accessed February 21, 2017)

[24] Moritz Oppenheim, "The Rothschilds at Yom Kippur," 1849 (whereabouts unknown), reproduced in Georg Heuberger and Anton Merk, eds., *Moritz Daniel Oppenheim: Jewish Identity in 19th Century Art*, exh.cat. Jewish Museum, Frankfurt am Main, (Cologne: Wienand, 1999), 9, 214, 367.

[25] Moritz Oppenheim, *The Eve of the Day of Atonement* in *Bilder aus dem altjüdischen Familienleben* (commentary by Leopold Stein) (Frankfurt: Keller, 1881), unnumbered page.

[26] The Minkowski is in the collection of the Tel Aviv Museum of Art. See Batsheva Goldman Ida, *Fragmented Mirror, Exhibition of Jewish Artists, Berlin, 1907*, exh.cat. (Tel Aviv: Tel Aviv Museum of Art, 2009), 39. The Pilichowski painting was exhibited at the same exhibition. See Goldman Ida, 63.

[27] Print by Abraham Neu, reproduced in Krautheimer, *Mittelalterliche Synagogen*, 105, fig. 22; photograph from Samuel Gruber's website at http://samgrubersjewishartmonuments.blogspot.co.il/2010/05/germany-arsonists-attack-medieval-worms.html. Accessed November 21, 2016.

[28] Jacob Moellin (ca. 1360/5–1427), *The Customs of the Maharil*, (Machon Yerushalyim: Jerusalem, 2005), "The Laws of the Eve of Yom Kippur," §13, 322 (Hebrew). These candles are mentioned, without any explanation, in Sidney Steiman, *Custom and Survival A Study of the Life and Work of Rabbi Jacob Molin (Moelln)* [sic],(New York: Bloch, 1963), 91. On the different ways to count the number of commandments that both men and women are obligated to perform, see Moshe (Murray Jay) Rosman, "A Proto-Feminist Challenge to the Polish Rabbinate in the Eighteenth Century: The Introduction to 'Tekhino Imohos" by Sarah Rebecca Rachel Leah Horowitz," in *The Scepter Shall Not Depart From Judah: Leadership, Rabbinate, and Community In Jewish History: Studies*

Presented to Professor Simon Schwarzfuchs, eds. Joseph R. Hacker and Yaron Harel (Jerusalem: The Bialik Institute, 2011), 314 (Hebrew).

[29] Eliyahu Kitov, *The Book of our Heritage*, (Jerusalem: Yad Eliyahu Kitov, 1976), 43 (Hebrew).

[30] *JPS Hebrew-English Tanakh* (Philadelphia: Jewish Publication Society, 1999), 1633.

[31] Rabbi Moshe Isserles (1520–1572), (The Rema) *Commentary on the Shulchan Aruch*, (Jerusalem: Ketuvim, 1993), "the Laws of Yom Kippur," §610; 179 (Hebrew).

[32] Shlomo Ashkenazy, *Generations in Judaism: Each Generation and its Sages* (Tel Aviv: Don Publishing, 1987), 322 (Hebrew).

[33] Some examples: Miriam Sperber, *From Berdichev to Jerusalem*, (Jerusalem: Private Press, 1981), 59 (Hebrew); Ashkenazy, *Generations in Judaism* ,323–324, (Hebrew).

[34] Pauline Wengeroff's memoirs were originally published in German as *Memoiren einer Grossmutter: Bilder aus der Kulturgeschichte der Juden Russlands im 19. Jahrhundert* (Berlin: Poppelauer, 1913). The two English translations of it are: Pauline Wengeroff, *Rememberings: the World of a Russian-Jewish Woman in the Nineteenth Century,* trans. Henny Wenkart (Bethesda: University Press of Maryland, 2000); *Memoirs of a Grandmother: Scenes from the Cultural History of the Jews of Russia in the Nineteenth Century,* trans. Shulamith Magnus (Stanford: Stanford University Press, 2010). Accounts of candle-making: Wenkart 55–56; Magnus, 159–160.

[35] Chava Weissler, "Measuring Graves, Laying Wicks," in *Judaism in Practice: From The Middle Ages Through The Early Modern Period,* ed. Lawrence Fine (Princeton: Princeton University Press, 2001) has translated one such text, 69–73.

[36] Weissler, "Measuring Graves, Laying Wicks," 61.

[37] William O'Dea, *The Social History of Lighting* (London: Routledge and Kegan Paul, 1958), 141–142.

[38] Herbert Thurston, "Candles," *The Catholic Encyclopedia*, Vol. 3 (New York: Appleton, 1908), 247.

[39] For a recent discussion of this see Roberta Rosenthal Kwall, *The Myth of the Cultural Jew* (Oxford: Oxford University Press, 2015), 55–60.

[40] *Babylonian Talmud*, Brachot 17a.

[41] For a discussion of this issue from a contemporary point of view, see Pamela Barmash, "Women and Mitzvot."
https://www.rabbinicalassembly.org/sites/default/files/public/halakhah/teshuvot/2011-2020/womenandhiyyuvfinal.pdf. Accessed November 3, 2016.

[42] For example, *Bereishit Rabba*, 17:8.

[43] Daniel C. Matt, trans., *The Zohar* (Stanford: Stanford University Press, 2004), Part 1, 48b, 266.

[44] Judith Baskin, ""She Extinguished the Light of the World": Justifications for Women's Disabilities in 'Abot de-Rabbi Nathan'," in *Current Trends in the Study of Midrash*, ed. Carol Bakhos (Leiden: Brill, 2006), 277–297.

[45] Shalom Sabar, "Childbirth and Magic," in *Cultures of the Jews: a New History*, ed. David Biale (New York: Schocken, 2002), 682. On the importance of light in the circumcision ceremony, see Baskin, "She Extinguished the Light of the World," 708.

[46] Chava Weissler, *Voices of The Matriarchs: Listening to The Prayers Of Early Modern Jewish Women* (Boston: Beacon Press, 1998).

[47] Nashman-Fraiman, "The Sabbath Lamp," 152–153.

[48] See Yael Levine, "Women who Composed Prayers for the Public: An Historical Study," in *Kenishta Studies of the Synagogue World* 2, ed. Joseph Tabory (Ramat Gan: Bar Ilan, 2003), 89–97 (Hebrew).

[49] Devra Kay, *Seder Tkhines*, (Philadelphia: Jewish Publication Society, 2004).

[50] Weissler, *Voices of The Matriarchs* ; Rosman, "A Proto-Feminist Challenge to the Polish Rabbinate in the Eighteenth Century."

[51] Weissler, *Voices of The Matriarchs*, 60; Sarah Bat Mordechai, *Tehinnah Shloshah She'arim* (Vilna: Fin, 1874), 7 (Yiddish).

[52] Reproduced in Grace Cohen Grossman, *Jewish Art* (Southport: Hugh Lauter Levin: 1995), 174-175.

[53] See Rosman, "A Proto-Feminist Challenge to the Polish Rabbinate in the Eighteenth Century, especially the conclusion, 315–316.

[54] Nashman-Fraiman, "The Sabbath Lamp," 146; Shalom Sabar, "'The Eternal Sabbath' – Electric Sabbath Candles. The History of a Folk Tradition from a Modernist Perspective," in *The Ethnographic Department of the Museum of the Contemporary*, eds. Lea Mauas and Diego Rotman (Jerusalem: Hearat Shulaym, 2017), forthcoming.

[55] The first reference appears in Exodus 20:7; the second in Deuteronomy 5:11.

[56] Nashman-Fraiman, "The Sabbath Lamp," 152–153.

[57] Responsa of Yehuda Tzvi Blum, *Sefer She'arit Yehuda* (Tel Aviv: Zohar Press, 1973), 23 (Hebrew); Rose Pesotta (1896–1965), *Days of Our Lives* (Boston: Excelsior, 1958), 87.

[58] Kwall, *The Myth of the Cultural Jew* refers to these practices as "bottom-up."

[59] Murray Jay Rosman, *How Jewish is Jewish history?* (Oxford: Littman Library of Jewish Civilization, 2009), 76.

CHAPTER SEVEN

IMAGERY OF LIGHT IN MEDIEVAL KABBALAH

SANDRA VALABREGUE-PERRY

Introduction

In this paper, I present a basic typology of imagery of light in thirteenth and fourteenth-century medieval theosophical Kabbalah. The topic is very broad, since light is the most important image that Jewish philosophy and mysticism use to describe metaphysical processes, like many other religious systems.[1] The wide use of such imagery is apparently due to light's abstraction and immateriality, but also, to a large extent, because of its power to give form and structure to the divine, as I shall demonstrate from its uses in the Jewish mystical tradition.

Theosophical Kabbalah, in its most well-known form, is based on a system of ten *sefirot*, or divine powers, with which God created and sustains the world. The word *sefirah* itself echoes the lighting of the sapphire from Ezekiel's vision.[2] Kabbalistic literature presents the *sefirot* in a large spectrum of light images, shades, and colours. Among the most important such images are the efflux of light; primordial light and primordial darkness; light as a point, a line or a circle; surrounding light; light that measures; light as prism or as mirror; and light in its various colours; candle light, coal light, and light as air or fire. Kabbalistic literature uses similar imagery to represent human and divine realms: light that is darkness for us, the light of the intellect, hidden light, and the light preserved for the righteous at the end of time. In this paper, I will present this rich imagery and focus on the mechanisms of light – *opticus aparitus*, to use Elliot Wolfson's term.[3] I will show how images of light structure the inner divine world, and explore what kind of divine space they construct and sustain.

This article is part of a larger ongoing study on the role of images in Jewish mysticism, and it should present some nascent results. Mystical language, whether secular or religious, aims to describe invisible phenomenon, and therefore makes use of paradoxical semantic statements. Thus, such language accomplishes the impossible by giving structure and model to what cannot be seen. A dominant theological reading of this language perceives the function of symbols, metaphors, and allegories as strictly descriptive. In such a reading, the images do not need to bear any resemblance with the concepts that they allegorically describe. This theological and philosophical method of reading imagery is problematic when applied to medieval Kabbalah, for those who use it tend to forget that the language of imagery is not merely a means to an end, a metaphor to explain some theoretical concept, but actually that the image fuses what is being represented and its representation in an astonishing way.

Unlike philosophical systems of thought, which limit themselves to speculating on what can be said about God with human language, medieval Kabbalah bypasses these epistemological limits by using images. Not only do images overcome linguistic limitations, but they are the main tool, if not the only one, for representing the un-presentable. In many aspects when it comes to theology, Kabbalah prefers images because of their power to create representation, more than their power to explain. I propose to address the images used in medieval Kabbalah not only as descriptive metaphors, but also as formative, and thus as founding and shaping the divine space.

This paper offers a short journey through a limited number of images of light in medieval Kabbalah to show how images are an example that can facilitate our understanding of a theological axiom and demonstrate that they embody the theosophical divine structure itself. The image *is* the theological axiom, rather than being just a specific metaphor for it. Theosophical images of light create a pattern, a structure, in a way that establishes the inner divine space, the *sefirotic* world. As we shall demonstrate, images of light give form to the kabbalistic conceptions of creation, emanation, and prophecy.

Conceptions of light and vision

Let us first open with a brief overview of the history of physical light.[4] One of the most ancient doctrines in this regard is the "extramission theory" of the atomists, which states that vision occurs by means of physical contact, when *eidola* (images) enter into the eye. Later came the "intromission theory," defended by Plato and Euclid, which understands vision as a kind of fire that goes out from the eye and meets an object. Aristotle was the first to write a systematic discussion of vision, which initiated a long tradition of adaption and combination of the intromission and the extramission theories. Aristotle criticized the two earlier theories, and directed his attention to the medium through which vision occurs, light, which is a state of transparency, through which, when actualized, the object becomes visible. From Greek to Arabic culture, optics reached new peaks, with Galen, Al-Kindi, and Al-Hazen.[5] The Christian Middle Ages, under the influence of Al-Kindi, also produced different theories that combined previous ones. Among the prominent scholars who studied this question were Vitello, Robert Grosseteste, and Roger Bacon.

Aside from these various theories of vision, medieval scholars also addressed the question of how light interacted with matter. According to their understanding, this interaction produced three kinds of light: direct, reflected, and refracted. Such a typology is found among Jewish scholars such as twelve and thirteen centuries Abraham bar Ḥiyya and Shem Tov ibn Falaquera, and later, in the sixteenth century, in the writings of Joseph Taitachek and his pupil Shlomo Alkabetz.[6]

Scholars were interested not only in the scientific aspects of light, but also in its direct connection to theology. In the Middle Ages, the desire to strengthen the analogy between optics and theology found its expression in the work of the prominent scholar Roger Bacon. More than anyone else, Bacon saw optics and epistemology as intertwined, for if physics and metaphysics go hand in hand, optics and epistemology open the possibility of reaching God. In his understanding, direct light, refracted light, and reflected light correspond to three degrees of knowledge: divine illumination, acquired knowledge, and empirical knowledge.[7] Direct light is the light of the enlightened one, and is the most perfect. Refracted light is knowledge, but reflected light is weak, for it is merely the light of the senses, which only stipulate whether things exist or not. However, there are some passages in Bacon’s work where this hierarchy between refracted and reflected lights is inverted. The reason for this change derives from Bacon's optical knowledge. This is a good example of the interaction between physics and metaphysics; the new scientific understanding of refracted light elevated its position relative to reflected light. Medieval science revealed that refracted light is not merely what is lost by the reverberation of light in water, but something else, a tool that makes it possible to see better. Using this knowledge, medieval scientists created the first individual lenses, such as binoculars and eyeglasses. Thus, the study of refracted light was instrumental both in mechanically improving vision and in changing theological assumptions.

Christian scholars such as Thomas Aquinas and Vitello tended toward an allegorical reading of the correspondence between light and God, where light is a metaphor for God. Nonetheless, some sources offer a more radical concept of total identification between light and God. This is the case, for example, in the *Liber de intelligentiis*, a widely disseminated medieval philosophical compilation in which divine light and natural light are said to be one and the same.[8]

God as Light and as Darkness

Much like Islamic and Christian sources and Jewish philosophy, medieval Kabbalah reflects the scientific knowledge of its time. It owes to Neoplatonic traditions the description of emanation in terms of light. However, Kabbalah not only offers a wider range of images, but also a more radical view. Like the *Liber de intelligentiis*, medieval Kabbalah does not tend towards an allegorical reading and seems to see the correspondence between God and light as literal. As we shall see, the light is the divine space itself. Even the stars (the sun and the moon) correspond to the *sefirot*. However, the main indication that Kabbalah is presenting a radical new system of imagery is not that it equates natural light with divine light, but that theosophical language, in contrast to philosophical argumentation, does not speculate about the process of emanation, but simply establishes this process by mean of images.[9] No scholastic demonstration or discussion can be found, because it is not an analytic approach, but rather an esoteric one that does not need to prove but only to display its knowledge when needed. The web of images projects directly into the divine realm and bypasses epistemological demonstrations.

A mode of thinking that works in images, rather than analysis, is usually omnivorous, using images from various sources. A brief survey of images of light in medieval Kabbalah will demonstrate such eclectic diversity and confirm that we are looking at a figurative, image-based system of thought. For example, we see that Kabbalistic literature uses images from the intromission theory of vision in this literature, even though this theory was already obsolete. The fourteenth-century commentary of Moshe Botaril on *Sefer Yeẓirah* compares the source and emanation of divine light to the light that beams out of the darkness of the pupil of the eye:

> This is what the Sages meant about the verse "*And the Lord passed by before him, and proclaimed*". He is one and united as the flame of fire that is unified in its hues as hinted in this verse "*that the Lord thy God, is a consuming fire*," and all the emanation from him is emanated as the light outgoing from the blackness of the eye.[10]

This image can already be found in early Kabbalah. For example, we see in the *Iyyun* literature, in the thirteenth century: “This light was generated from His innermost, like the light of the eye, which issues forth from the pupil of the eye, from which the light sparkles.”[11] This text uses natural optics to explain theological optics, in order to show that the two are identical, and thus both truth. Jewish philosophical sources, too, discuss the interaction between light and darkness. For example, Isaac Israeli explains that human eyes, which have eyelids, can see colours, whereas reptiles’ eyes, which have no eyelids, cannot; this is because the eyes need to be able to close in order to recharge in the darkness, to increase their inner light, in order to be able to see better. To perceive colours, one needs to have an eyelid, because only closing the eyes can enable them to collect the necessary inner light.[12] For reptiles, which have no eyelids, the light that comes out of the eye is weakened and confused by the light coming from the sun, and therefore they cannot see colours.[13]

The Kabbalist Azriel of Gerona presents a different example, in which he draws on empirical observations, of the interaction between darkness and light. According to him, darkness is necessary not only for the eye to recharge, but also for natural light to be able to shine. He writes that a candle is lit from the darkness that surrounds it.[14] If, on an epistemological level, God is light and absence of God is darkness, on a theological level darkness has an ontological status: only out of it can light be created. Most of the time, Kabbalists address God not only as primordial light, but also as primordial darkness; only rarely do they view primordial darkness as a negative force of impurity.[15]

Primordial light[16] and primordial darkness[17] are thus two aspects of the same concept; they both designate the origin of everything, with light being the revealed aspect and darkness being the concealed one. In different traditions, and sometimes even within the same tradition, there are different understandings of darkness. In the philosophical tradition, darkness is the absence of light; but in Kabbalistic texts it is understood also as a greater light, a light that is so strong that it is darkness for us. This view is not only a way to give form to an invisible light, but also to precede light, as radiance. At the beginning there was nothing, nothing to be seen. This nothingness is obscurity, yet the nothingness is also the primordial light. This idea also articulates another postulate: while Man can only create light out of another light, like that of a candle, God can create something out of nothing, light out of darkness. Following Todros Abulafia description:

> Know that the night preceded the day, for it is written: *"And there was evening and there was morning*". Therefore, the obscurity preceded the

> light, for it is first written *"darkness was upon the face of the deep"*, and then *"let there be light…"*
>
> Man lights up a candle from another candle, [… whereas] God brought out the light from within the darkness.[18]

This is where God's visibility becomes deeply connected to yet another visibility in potential, that of His creation and His actualization. The process of creation is the process of bringing into visibility, from obscurity into light. Within the same conceptual web of darkness is the concept of concealment. Concealment and expansion, transcendence and immanence are the two poles of the same very axis between primordial darkness and light.

It is commonly understood that if one looks directly at the sun, the source of light, it will blind and darken one's vision. Similarly, in Neoplatonic philosophy and kabbalistic literature, the source of light, which is the source of creation, must be concealed in the creation. The Zoharic literature describes two steps in the creation process as the simultaneous concealment and exposure of light: the preservation of the primordial light for the distant future, and the covering of light with garments.[19]

These different images of light understood as darkness – darkness as a primordial state of light, the darkness from which colours are seen, and the darkness that lights up and nourishes the light – are theoretical tools for understanding and articulating the beginning and origin of creation.

Light as Emanation

As we have seen, the image of darkness as a primordial state of light relates to questions of creation and beginning through relationships between opposites (light and darkness). Now let us look at the images of light as emanation, and the forms that they give to the relationship between God and the world, mainly between God and the *sefirot*. If darkness is equivalent to concealment, then light rays are equivalent to emanation. Unlike the philosophical and theosophical traditions that stress the ontological distinction between creator and creation, the images of the source of emanation and its emanation are meant to do the opposite: to depict an ontological continuity.[20] These images fall into two categories: those that describe emanation with reference to its source, and those that describe emanation with reference to its expansion and reception.

Let us begin with the first category, images that relate to the source of the emanation. The association between images of light and the theory of emanation goes back to the fourth-century philosopher Plotinus, who explained the theory of emanation with the image of the sun and its rays.

He writes: "One is the light before the light, an eternal irradiation resting upon the Intellectual Realm; this, not identical with its source, is yet not severed from it".[21] Medieval Arab and Jewish philosophers embraced metaphors of light throughout their writings. For Isaac Israeli, following his teacher Ibn Ḥasdai, the intellect comes into being from the conjunction of primordial matter and form. Next, a brilliant flame and light, from which arises the rational soul, is emanated. In a famous parable, he compares the intellect to the sun, as it shines through the glass of a window into a house. The glass, also referred to as "mirror" or "crystal," represents the rational soul, which needs to learn and memorize in order to actualize the light of the intellect. Only the light of the intellect can save the intellect from "the shade, cloud and mist of darkness . . . Hence the rational soul requires learning and memory, while the intellect is never affected by ignorance, because nothing intervenes between it and the power."[22] If pure and unmixed light is the power at the origin of everything, the world, in return, is the opacity and the density that result from being farther away from the primordial light. Solomon Ibn Gabirol, known for his influence on early Kabbalah, also uses the image of light to explain the conception of emanation.[23] Unlike Ibn Gabirol, Abraham ibn Ezra clearly opts for the extramission theory.[24]

In the literature of the Zohar, the source of light is said to be concealed, and can only be seen through the crystal, where the inner lights will be revealed:[25]

> Come and see: there are four lights. Three of them are concealed, and one revealed. A shining light. A light of radiance, shining like the heavens in purity. A light of purple, absorbing all lights. A light that does not shine but gazes toward these, receiving them, and they appear in it as a crystal (עשישתה) facing the sun. These three that we have mentioned are concealed and stand above this one, which is revealed.[26]

The paradigm of the sun (source of light), light, and mirror helps to overcome a major ontological obstacle, namely, how to explain the possibility of change and contact between the creator and its creation. In line with the philosophical sources quoted above, Isaac Ibn Latif writes: "All beings flow from Him and exist through Him in the manner of emanation and evolvement, and nothing is outside of Him, as the light emanates from the sun, and yet the light is nothing but the sun."[27] This image helps to explain how the light can be sent forth and yet still be the source of the light itself.[28] In the words of the author of *Tikkunei Zohar*: "When He created the world with the attribute of *reshit* (beginning), He descended upon it, but there was no diminution above, and similarly with every *sefirah*; like someone lighting a candle from a candle, where there is no diminution in the first one, nor in the second, *ad infinitum*, without end."[29]

Concealment of Light

As we have seen, the image of light is of great value in explaining changes as well as continuity. Change and continuity comprise the basic structure of the divine space, and images of light stress its omnipotence, transparency, and therefore its immediacy of emission. Transparency and immediacy are the signs of its divine origin, since it is beyond matter and form. However, since creation and emanation concern the way in which the universe can issue from the One, this first step still needs to be accounted for. Concealment is the key image used to explain this first step of creation. Concealment is related to the famous Kabbalistic thesis of *ẓimẓum*, because concealment is also the retreat of the light from its overwhelming brightness.[30]

In an early Kabbalistic text by Ezra of Gerona, his commentary on the Song of Songs, he explains concealment as a primordial matter in the form of a garment,[31] a conception that would find many echoes in other early Kabbalah texts and beyond: "You . . . are wrapped in a robe of light" corresponds to the light created on the first day of creation. That is, the emanation of light from Nothing."[32] In the same text, he writes:

> Let there be light [Gen 1:3]. Its meaning is that Divinity enwrapped Itself in that light as we have explained, meaning that though the amplification of that luminescence, the light entered into a state of actualization . . . it is for this reason that scripture states, "*it was good*," for the light was kindled candle from candle, granting the primal essence the power to expand and flow forth.[33]

The imagery of garments of light used to explain God's Creation as an act of creation out of primordial matter, goes back to ancient Jewish midrash, and echoes Plato's conception of primordial matter in the *Timaeus*. Alexander Altmann has stressed the importance of this imagery in medieval thought.[34] Interestingly enough, garments as primordial matter can also be understood as primordial form.

Light is not only seen as the source of everything, but also as giving form to everything. The very particularity of light lies in its spiritual presence and formative power. As Jacob ben Sheshet describes in his book, light is form without matter.[35] The aspect of light as having the power to give form is the source of many poetic images in Kabbalah, particularly in the literature of the Zohar. The Zoharic section called *kav ha-middah* (measuring line) portrays light as that which measures and gives dimensions to the *sefirot* and to the world:

> *Kav ha-middah*: there is length and breadth in the measuring, depth and height in the measuring, circle and rectangle in the measuring – the measuring line encompasses all. All the lights and all the mysteries neither ascend, nor descend, nor extend themselves, except by means of this measuring line, apart from the veritably uppermost regions [*En-sof* and *Keter*] where there is no measure…[36]

The image of the line should be considered among other geometrical images in which light takes form, such as the point and the circle, which are understood both as tools of creation and as the different stages of emanation. The point is one of the most important images of beginning, offering the possibility of concealment and expansion.

It is not by chance that the writings of theosophical Kabbalah offer no clear theory of its philosophical notions regarding creation, emanation, and the first matter and form. Although the Kabbalists have taken these notions from philosophical terminology, they are not interested in the analytical philosophical discussions from which the notions arose. Even in the field of Jewish philosophy, scholars are still discussing the meaning of these images and the theories behind them. Did Solomon Ibn Gabirol or Isaac Israeli have a concept of creation *ex nihilo*, or of primordial matter? Which came first, matter or form? Were they emanated simultaneously? These questions still do not have a definitive answer.[37] This is also the case with the kabbalistic texts. However, unlike the philosophers, from the outset kabbalists did not feel compelled towards the same kind of coherence or consistency. Their allegiance to an image oriented form of thinking gives them freedom in their use of philosophical notions, regardless of their occasional contradictions. They can accept a theory of creation *ex nihilo* if they understand darkness and light as total opposites, and then they may understand darkness as concealed light, and move over to a theory of creation through emanation. Similarly, they can adhere to a conception of light as primordial matter or as primordial form regardless of hierarchy. This is exactly the kind of flexibility that is allowed by thinking in the language of images, and one of its virtues is that it is comprehensive rather than analytical, fluid and associative rather than logical and static.

Light as Human Reflection

Although we have focused so far on images of light in descriptions of the process of emanation, Kabbalists also use such images in discussions of man's interaction with God. Light is the very blessing that is bestowed to a man of God, for the righteous or the mystics are said to be enlightened. If optics is the physical study of light, prophecy is its metaphysical laboratory, to which only a few have access. The name of the Zohar, which means *radiance* alludes to this. As it says in this work: "It is written: The enlightened will shine like the radiance of the sky and those who turn many to righteousness, like the stars forever and ever (Daniel 12:3). The enlightened will shine. Who are the enlightened? The wise one who contemplates, on his own, things that human beings cannot utter. They are called enlightened."[38] The enlightened are the ones who can receive direct light, that is, enlightenment that is the result of non-mediated knowledge. Through contemplation, the righteous access knowledge that the prophet would access through revelation; in both cases, they experience light. The association between divine light and prophecy is based on a Talmudic statement: "All the prophets gazed through an opaque glass [an *aspaklaria* that does not shine], whereas Moses our master gazed through a translucent glass [an *aspaklaria* that shines]".[39]

The model of these two glasses refers in Kabbalah to two kinds of prophecy. The first, Moses's prophecy, is experienced as a direct light, pure enlightenment, without parable or metaphor, as the mirror used by women that reflect an exact image.[40] The mirror has very often been presented as a model to explain God's creation, understood as the instantaneous reflection of the holder in the mirror.[41] The second mirror, the prophecy of the other prophets, is less refined, and reflects only blurred images.[42] The quality of the mirror depends on the quality of the light that is reflected, and sometimes, only the first mirror is described as reflecting pure light at all, whereas the second is described as reflecting vision[43] or as encompassing all the colours.[44]

This model of the two mirrors with different capacities of reflection, is used in Kabbalistic literature not only to explain levels of prophecy, but also, as in Zoharic literature, to explain the relationship between God and the *Shekhinah*, that is, between the *sefirot Tiferet* and *Malkhut*.[45] It is important to note that the relationship between God and the *Shekhinah* is also portrayed as equivalent to the flux of light and its reflection/absorption in early Kabbalistic literature, which uses the imagery of natural light and the mirror that shines.[46] This is in line with the correspondence between *Tiferet* and the sun, on the one hand, and *Malkhut*, the mirror that does not shine, and the moon.[47] The mirror that does not shine, like the moon, does not have a light of its own, and only reflects the source of light, the sun. Thus, the *sefirotic* structure represents the relationship between God and His reception in the world as the relationship between the sources of light, on the one hand, and a body that can only capture light, on the other – like the moon and sun in the physical world.

In this paper, I have attempted to show briefly how images of light not only helped Kabbalists give form to major theological questions, but also bestowed upon Kabbalah essential theological tools. The images of light as primordial form or matter, primordial light or darkness, as emanation, as enlightening or reflecting light, created the space and type of relationships where God, the world, and man could meet.

It does not do justice to the suggestive power of these images to view them simply as a metaphorical language to distance the divine realm from the human. Images are not only a metaphor to describe what cannot be described; rather, they serve as a metalanguage, and as a substitute to what cannot be seen. They are not merely a means to portray an unfathomable idea; but create a platform, a divine space where what has no form can be established. Within this world, images of light provide the most inner dynamism: creation and beginning, flow, expansion, sustention and conservation.

Notes

[1] Few works have been dedicated to the role of images and imagination in Kabbalah, but those that have been written are important. See Marla Segol, *Word and Image in Medieval Kabbalah* (New York: Palgrave, 2012); Daniel Abrams, "Kabbalistic Paratext," *Kabbalah* 26 (2012): 7–24; idem., "New Study Tools From the Kabbalists of Today: Toward an Appreciation of the History and Role of Collectanea, Paraphrases and Graphic Representations in Kabbalistic Literature," *Journal pour les études de la Cabale: Revue interdisciplinaire de recherche et d'information* 1 (1997): 1–7; Moshe Halbertal, "Of Pictures and Words: Visual and Verbal Representation of God", in *The Divine Image: Depicting God in Jewish and Israeli Art,* eds. Sharon and Ronit Sorek Weisner-Ferguson (Jerusalem, The Israel Museum, 2006), 7–13; Guilo Busi, *Qabbalah Visiva* (Torino: Einaudi, 2005); Haviva Pedaya, *Vision and Speech. Models of Revelatory Experience in Jewish Mysticism* (Los Angeles: Cherub Press, 2002) (Hebrew); idem., "Drawing and Picture in Kabbalistic Interpretation of Nahmanides," *Mahanayim* 6 (1994): 114–123 (Hebrew); Elliot Wolfson, *Through a Speculum that Shines: Vision and Imagination in Medieval Jewish Mysticism* (Princeton: Princeton University Press, 1994); idem., "Iconicity of the Text: Reification of the Torah and the Idolatrous Impulse of the Zoharic Kabbalah," *Jewish Studies Quarterly* 11, 3 (2004): 215–242. See also Katrin Kogman-Appel, "The Role of the Hebrew Letter in Making

the Invisible Visible," in *Sign and Design: Script as Image in a Cross-Cultural Perspective (300-1600 CE)*, eds. Brigitte M. Bedos-Rezak and Jeffrey F. Hamburger (Washington, D.C: Dumbarton Oaks, 2012), 153–166; and, more specifically on images of lights, Wolfson, *Through a Speculum that Shines*, 270–288; idem., "Hermeneutics of Light in Medieval Kabbalah" in *The Presence of Light: Divine Radiance and Religious Experience*, ed. Matthew T. Kapstein (Chicago: University of Chicago Press, 2004).

[2] Ezekiel 1:26. See also the vision of the seventy elders of Israel in Exodus 24:10.

[3] Wolfson, *Through a Speculum that Shines*, 274.

[4] Vaco Ronchi, *The Nature of Light*, (Harvard: Harvard University Press, 1970); Lindberg Carter, *Theories of Vision from Al-Kindi to Kepler* (Chicago and London: University of Chicago Press, 1976).

[5] On the influence of Algazali and the metaphysics of light on early Kabbalah see Marc Verman, *The Books of Contemplation, Medieval Jewish Mystical Sources* (Albany: 1992), 130. On lights in the Iyyun literature, see Gershom Scholem, *Origins of Kabbalah*, (Princeton; Princeton University Press, 1987), 333–336.

[6] Bracha Sack, "the Simile of the Three Lights in Solomon Alqabes' *'Ayeleth 'ahavim* ," in *Studies in Jewish Religious and Intellectual History, Presented to Alexander Altmann on the Occasion of His Seventieth Birthday*, eds. Siegfried Stein and Raphael Loewe (London: University of Alabama Press, 1979), 43–63. Apparently medieval Kabbalah and the Kabbalah after Cordovero prefer the two *Aspaklaria* models rooted in the midrash images over scientific models such as the three lights model.

[7] David C. Lindberg, *Roger Bacon and the Origins of Perspectiva in the Middle Ages* (Oxford: Clarendon Press, 1996), 327–9.

[8] Pierre Duhem, *Le système du monde: histoire des doctrines cosmologiques de Platon à Copernic* vol. V (Paris: Librairie Scientifique Hermann, 1913–1959), 370–372.

[9] See for example the fourteenth century philosophical discussion of R. Judah Romano in Joseph Sermonetta, "Light – its signification and role in the Exegesis of *Bereshit*," *Massu'ot, Studies in Kabbalistic Literature and Jewish Philosophy in Memory of Prof. Ephraim Gottlieb*, eds. Michal Oron, Amos Goldreich (Jerusalem: Bialik Institute, 1994), 343–360 (Hebrew).

[10] *Sefer Yeẓirah* (Jerusalem: Yeshivat Kol Yehuda, 1988), 35.

[11] Marc Verman, *The Books of Contemplation, Medieval Jewish Mystical Sources* (Albany: State University of New York Press, 1992), 78; Elliot Wolfson, "Hermeneutics of Light in Medieval Kabbalah," in *The Presence of Light: Divine Radiance and Religious Experience*, ed. Matthew T. Kapstein (Chicago :University of Chicago Press, 2004), 108–112.

[12] For a kabbalistic technical explanation of colour vision in the innermost part of the closed eye, see Wolfson, *Through a Speculum that Shines*, 380–383.

[13] Alexander Altmann and Samuel M. Stern, *Isaac Israeli: A Neoplatonic Philosopher of the Early Tenth Century* (London: Oxford University Press, 1958), 94.

[14] R. Azriel of Gerona, *The Commentary on the Agadot*, ed. Isaiah Tishby (Jerusalem: Akademon, 1945), 86 (Hebrew).

[15] Joseph of Hamadan, *Fragment of a Commentary on Genesis*, ed. Charles Mopsik (Paris: Verdier, 1998), 16-17 (Hebrew). On the idea that shadow is an emanation of the left side, see Moshe Idel, "Kabbalistic Material from R. David ben Yehuda he-Hassid's School" (Hebrew), *Jerusalem Studies in Jewish Thought* 2 (1983): 183.

[16] As Todros ben Joseph Abulafia writes in *Sha'ar ha-Razim*, ed. Michal Kushner-Oron (Jerusalem: Bialik, 1989), 121: "[The first emanation, *Keter*] is emanated from the first cause which is drawn from the inner and primordial light, the root of the root."

[17] On the different kind of obscurity attributed to *Keter*, *Gevurah* or *Malchut*, see Michal Oron, *Sefer ha-Shem attributed to R. Moses de Leon* (Los Angeles: Cherub Press, 2010), 183, n. 831; Verman, *The Books of Contemplation*, 58, 149.

[18] Abulafia, *Sha'ar ha-Razim*, 62.

[19] See note 31, below.

[20] See Sandra Valabregue, "Philosophy, Heresy and Kabbalah's Counter-Theology" *Harvard Theological Review* 109:2 (2016): 233–256.

[21] Plotinus, *Enneades* V3:12, 15. On the importance of Plotinus's light metaphysics for medieval Neoplatonism, see Bogdan Gabriel Bucur, "'The Place of Splendor and Light': Observation on the Paraphrasing of Enn. 4.8.1 in the Theology of Aristotle," *Le Muséon* 119 (2006): 271–292.

[22] Altmann and Stern, *Isaac Israeli*, 99.

[23] Vincent Cantarino, "Ibn Gabirol's Metaphysics of Light," *Studia Islamica* 26 (1967): 49–71. Sarah Pessin, *Ibn Gabirol's Theology of Desire: Matter and Method in Jewish Medieval Neoplatonism* (New York: Cambridge University Press, 2013); Jacques Schlanger, *La philosophie de Salomon Ibn Gabirol, Étude d'un néoplatonisme* (Leiden: E. J. Brill, 1968), 47–51.

[24] Shlomo Pines, "'He Called Forth to the Nothing and It Split': Research on the Keter Malkhut of Solomon ibn Gabirol," *Tarbiz* 50 (1982): 345, n. 24 (Hebrew) notes that Arabic philosophers attributed this theory to Plato; Israel Efros, *Studies in Medieval Jewish Philosophy* (New York: Columbia University Press, 1974), 239–241.

[25] Yehuda Liebes, *Section of the Zohar Lexicon* (PhD diss., Hebrew University, 1976): 134–167 (Hebrew).

[26] Zohar 2:23a; *The Zohar*, trans. Daniel Matt, vol. 4 (Stanford: Pritzker, 2007), 80. See also "Isaac of Acre's Commentary on Sefer Yetzirah," *Kiryat Sefer* 31 (1956): 383 (Hebrew).

[27] Isaac ibn Latif, *The Gate of Heaven* vol. 1, ed. Raphael Kohen (Jerusalem, 2015), 484.

[28] This is similar to the idea of immediacy, i.e., that creation is out of time, and therefore there is no time delay between the source of light and its emanation, just as a shadow appears as soon as light is shined on an object. See note 17. above.

[29] Tikkunei Zohar, 40b; Isaiah Tishby, *The Wisdom of the Zohar,* 3 vols, trans. David Goldstein. (Oxford: Littman, 1989), 1: 274.

[30] Moshe Idel, "The History of the Concept 'Zimzum' in Kabbalah and in the Research," *Jerusalem Studies in Jewish Thought* 10 (1992): 59–122 (Hebrew).
[31] "The Lord Blessed Be He, created a primordial light which enlightened until the world could not suffer it any more. What did the Lord do? He made a light for the light, to dress one with the other – and then for all the other lights, until all the worlds could be sustained in their existence and suffer (the light)", Zohar 3:204b.
[32] Rabbi Ezra ben Solomon of Gerona, *Commentary on the Song of the Songs*, trans. Seth Brody (Kalamazoo: Western Michigan University, 1999), 109.
[33] Ezra of Gerona, *Commentary on the Song of the Song*, 112; Shlomo Ibn Gabirol, *Fons Vitae* IV: 14.
[34] Alexander Altmann, *Studies in Religious Philosophy and Mysticism* (Ithaca: Routledge, 1969), 128–139. For a discussion of how Kabbalists reinterpreted *creatio ex nihilo* as an emanation theory, see Valabregue, *Philosophy, Heresy and Kabbalah's Counter-Theology*, 245-25. For a study on garments in the Zohar, see Dorit Cohen-Alloro, *Secret of the Garment in the Zohar* (Jerusalem: Hebrew University, 1987), 12–17; 68–89 (Hebrew).
[35] Jacob ben Sheshet, *Sefer meshiv devarim nekhoḥim*, ed. Efraim Gottlieb (Jerusalem: National Academy of Sciences, 1968), 193,120 (Hebrew).
[36] *Zohar Hadash* 57a.; Tishby, *Wisdom of the Zohar*, vol. 1, 271–72.
[37] Sarah Pessin summarizes and presents her own position on the subject: "Jewish Neoplatonism: Being above Being and Divine Emanation in Solomon Ibn Gabirol and Isaac Israeli," in *The Cambridge Companion to Medieval Jewish Philosophy*, eds. Daniel Frank and Oliver Leaman (Cambridge: Cambridge University Press, 2003), 91–110.
[38] Zohar 2:23a (*The Zohar*, trans. Matt, vol. 4, 79); Wolfson, *Through a Speculum That Shines*, 383–392.
[39] TB Yebamot 48b. There is a Christian parallel in I Corinthians 13:12: "Now we see dimly through a mirror, but in glory face to face." See Immanuel Lev, "Aspaklaria: A Chapter in Mineralogy of the Jews," *Festschrift für Professor Samuel Krauss, anlässlich seines 70. Geburtstages* (Jerusalem: Mass, 1937), 10–14 (Hebrew); Joshua Brand, "Aspaklaria," in *Benjamin de Vries Memorial Book*, ed. Ezra Melamed (Tel Aviv: Tel Aviv University, 1969), 110–118 (Hebrew).
[40] Oron, *Sefer ha-Shem*, 126.
[41] This image in found in Solomon Ibn Gabirol's *Fons Vitae* (V, 37) and in Azriel of Gerona; Gershom Scholem, *Studies in Kabbalah,* 1 (Tel Aviv: Am Oved, 1998), 57 (Hebrew). See Sarah Heller-Willensky, "Isaac ibn Latif, Philosopher or Kabbalist?," *Jewish Medieval and Renaissance Studies* (1967): 209; Oron, *Sefer ha-Shem*, 185.
[42] Oron, *Sefer ha-Shem*, 129; Abulafia, *Sha'ar ha-Razim*, 50.
[43] For example, in comparison to the pure light experience of Moshe, the others are made of images, like a dream vision. *R. Moshe de Leon's Sefer Sheqel ha-Qodesh*, ed. Charles Mopsik (Los Angeles: Cherub, 1996), 69 (Hebrew).
[44] Moshe *Halamish, Commentary of Rabbi Yoseph ben Shalom Ashkenazi* on *Genesis Rabbah* (Jerusalem: Magnes, 1985), 43 (Hebrew). For the interpretation of

the philosophical concept of form as colour, see: Gershom Scholem, *Kitvey Yad be-Kabbalah* (Jerusalem: Hebrew University Press, 1930), 58 (Hebrew).

[45] Daniel Abrams, "“A Light of Her Own”: Minor Kabbalistic Traditions on the Ontology of the Divine Feminine," *Kabbalah* 15 (2006): 7–29. I wish to thank Ayelet Naeh for sharing her research on the *aspaklaria* in the Zohar. See Ayelet Naeh, "I Which Is Not One: On the Relation to Oneself in Kabbalah Language" (Hebrew), forthcoming.

[46] See for example: Sheshet, *Sefer meshiv devarim nekhoḥim*, 128; Azriel of Gerona, *Perush ha-Agadot*, 53.

[47] Moshe Halbertal, *By Way of Truth: Nahmanides and the Creation of Tradition* (Jerusalem: Shalom Hartman Institute, 2006), 211–181 (Hebrew); Haviva Pedaya, *Nahmanides: Cyclical Time and Holy Text* (Tel Aviv: Am Oved, 2003), 379–359 (Hebrew).

PART 4:

CONCEPTUAL FUNCTIONS OF LIGHT

CHAPTER EIGHT

THE LADDER OF LIGHT IS THE LADDER OF MONEY

YAIR BARAK

St. Bonaventure was a theologian-philosopher, a man of his time whose writings and religious career exemplified the intellectual life of the contemporary educated elite in thirteenth-century Europe. His writings combine the Neoplatonist Augustinian tradition and rediscovered Aristotelian philosophy. His metaphysics formulates a universe of which light is one of two fundamental ontological components – *materia* as universal matter and light as a universal form.

Bonaventure's philosophy is a brave effort to bridge all the gaps and contradictions of his time: Plato versus Aristotle; science versus tradition; commercial revolution that led to economic boom versus the Franciscan ideology that extolled and demanded poverty; and a society in transition that sought order and organization by simultaneously establishing two different and mutually contradictory institutions, the university (ca. 1200) and the Inquisition (1232).

Though Bonaventure preached pauperism, he lived in Paris, which was a flourishing monetary centre, and hence, as both a teacher at the University Paris and head of the Franciscan Order, which owned huge fiefs, he was directly exposed to the monetary revolution that was taking place. His anti-property ideology compelled him to understand and define material property and *pecunia* (currency).

Aware of the worldly world and the new function of coins as a quantifying, continuous scale of all material things and services, he turned light into a quantifying measure of the distance to God – a new Jacob's ladder from heaven to earth, but made of light.

Light constitutes the hierarchy among all beings – physical and abstract

St. Bonaventure opened his well-known work *Itinerarium Mentis in Deum* (The Journey of the Mind to God) with the sentence: "As I begin, I call upon the first beginning, the father of lights from whom all illumination descends, from whom every good gift and every perfect gift derives."

According to the epistemology of St. Bonaventure, the world, with all its objects, has two basic elements: hyle (matter), the *materia universalis* – universal material or substance – and the universal form. Universal material is uniform and common to all objects. The universal form, which distinguishes between objects, is light.

Bonaventure accepted Aristotle's basic distinction between matter (substance) and form and applied it to all beings – corporeal and abstract – and by doing so, constructed a unified world – a monistic world – as opposed to a dualistic one that distinguishes between materialistic phenomena and the spiritual. This dichotomy is, fundamentally, Plato's epistemology (from his perspective it is an ontology), according to which the real world is the world of the abstract and pure forms, the realm of ideas, of which the corporeal world is merely an imitation – a mimesis. Hyle is uniform, like the identical atoms that constitute molecules, and it is a universal ground, while form differentiates objects from one another. Matter is passive, while form is active. While hyle is the *materia universalis*, light, according to Bonaventure's innovative idea, is the universal form. According to Frederick Copleston:

> [Light] is not, properly speaking, a body but the form of a body, the first substantial form, common to all bodies and principle of their activity, and the different kinds of body form a graded hierarchy according as they participate more or less in the form of light. Thus the 'empyrean' stands at one end of the scale, while the earth stands at the other, the lower end. In this way the light-theme, so dear to the Augustinian School and going back to Plato's comparison of the Idea of Good with the sun, finds a prominent place in the philosophy of St. Bonaventure.[1]

The quantity of light determines the hierarchy of all beings – corporeal and abstract – on the same scale. John F. Quinn explains why Bonaventure identified light as a universal measure:

> Light is a substantial form . . . since it is the noblest of corporeal forms . . . The nobility of the light comes from its priority in *being* and in time over other corporeal forms. As a consequence, the bodies formed after the

> production of light have a greater and lesser dignity in substantial *being*, and specific difference, according as they participate in light.[2]

Light is accepted as the noble omni-measure of all things in the universe, including God. Light is the first creation, and its simultaneous function is both to light and to grade and determine the hierarchy of all beings as they participate more or less in the form of light. The amount of light in each being situates it within the hierarchy of beings, and since God is the absolute light, the supreme light, this hierarchy is a measure of distance from God. Thus, more light means more proximity to God.

Even if a human being is endowed with a great amount of light, he or she is able to move closer to God through a ladder of deeds and acts that contain an ever- increasing measure of light. The first step on the ladder of ascent toward God is prayer: "By so praying, we are given light to discern the steps of the soul's ascent to God,"[3] Bonaventure wrote.

The four sources of Bonaventure's theology-philosophy

Bonaventure's philosophical and theological discourse was based upon attributing to light an original ontological status as a means of universal measurement. It was inspired by four sources. The first of these is the universal status of light in the monotheistic religions, while the other three are anchored in the context of his time: the physical and philosophical discourse on light; the spread of monetary culture in the economy of western Europe, which was transformed from a barter economy to one based on currency – a monetary economy; and the passion for measurement and quantification that were responsible for the monetary revolution in the epistemology of the intellectuals of his era.

While the discourse on light took place within the republic of the learned, the monetary culture was being practiced in all the spheres of social and religious life to which Bonaventure was exposed. Bonaventure was a monk of the Franciscan mendicant order, but his life proceeded within the secular and commercial world outside the monastery – at the University of Paris and as head of the Franciscan Order (starting in 1257). At that time, Paris was a financial and monetary centre. In his role as head of the monastery, he travelled on foot (never riding on the back of any animal) throughout the continent and was thus exposed to the monetary reality of his era. This contact with the developing commerce of Europe demanded his practical response as the head of a mendicant monks' order and his intellectual engagement as a leader of an ideological system that advocated intentional poverty and opposed the individual acquisition of private property (although the monastic order itself owned property).

The first source: The supreme status of light in the monotheistic religions

The Bible frequently identifies light with God, for example in the line "The Lord is my light and my salvation" (Psalms 27:1). In Sura 24 of the Qur'an, called "The Light" (An-Nur), we find "Allah is the light of the heavens and the earth . . . And he whom Allah has not granted light, for him there is no light." In the Mass, the central Christian prayer, we find following: "*Deum de deo/Lumen de lumine*," which means "God is light." The parallelistic trope equates God with light.

In accordance with these perceptions, absolute light is God. He is the absolute spirit – the highest extreme of the hierarchy of the *ontos* – of being. According to Quinn, "[i]n Bonaventure's view, light has a property: the spiritual nature found principally in the Creator, who is purely light and, as such, the cause of the light in the human soul. The light of truth in the soul is referred analogically to the light of truth in God, as an effect is referred to its cause."[4]

According to Bonaventure, the primacy of light and its independence of any material substance is confirmed in the account of Creation; the appearance of light is the first act of creation (Genesis, 1: 3), whereas the sun, ostensibly the source of the light, was created only on the fourth day, along with the other illuminators (Genesis, 1: 16). Thus, light preceded the sun, both conceptually and in time. Indeed, as Robert Lopez has noted, "[t]he Augustinian theory of divine illumination as the source of all human knowledge never lost its appeal. St. Bonaventure made it the starting point for his own pathway to God through a progression of divine illumination."[5] Lopez connects Bonaventure to the traditional Christian illumination chain and notes that "the theories of the Oxford School of philosophers were linked with it too, but from a different angle, for their objective was not so much divine wisdom as divine power."[6]

The second source: The scientific and philosophical discourse on light

In the studies of light of Bonaventure's contemporaries, members of the Oxford School, Roger Bacon and Robert Grosseteste, we can discern the beginnings of an examination of light by means of mathematical and physical tools in an attempt to understand the phenomenon in terms of material and expansion (Grosseteste) or through geometric and mathematical concepts (Bacon).

Grosseteste, who wrote his treatise *De Luce* (On Light or The Beginning of Forms) from 1225 to 1230, was close to the Franciscan Order, which was, as mentioned above, headed by Bonaventure from 1257. Hence, it is quite likely that Bonaventure was acquainted with Grosseteste's treatise. This acquaintance is evident in the first line of *De Luce*: "The first corporeal form which some call corporeity is in my opinion light."[7] He explains the physical reason for this: "For light of its very nature diffuses itself in every direction in such a way that a point of light will produce instantaneously a sphere of light of any size whatsoever, unless some opaque object stands in the way."[8] Grosseteste continues the physical discourse, aiming to support with rational and scientific reasoning the unique status of light in nature:

> I have proposed that it is light which possesses of its very nature the function of multiplying itself and diffusing itself instantaneously in all directions . . . Corporeity, therefore, is either light itself or the agent which performs the aforementioned operation and introduces dimensions into matter in virtue of its participation in light, and acts through the power of this same light. But the first form cannot introduce dimensions into matter through the power of a subsequent form. Therefore, light is not a form subsequent to corporeity, but corporeity itself.[9]

Prima facie, Grosseteste's physical arguments contradict the common religious belief that light is above the whole world since light is God. But his next statement modifies the status of light and makes it a base for Bonaventure's notions:

> Furthermore, the first corporeal form is, in the opinion of the philosophers, more exalted and of a nobler and more excellent essence than all the forms that come after it. It bears, also, a closer resemblance to forms that exist apart from matter. But light is more exalted and of a nobler and more excellent essence than all corporeal things. It has, moreover, greater similarity than all bodies to the forms that exist apart from matter, namely, the intelligences. Light therefore is the first corporeal form.[10]

Thus, Grosseteste's physical claims are consistent with Bonaventure's metaphysical notions regarding the supreme status of light as a universal form. It is also clear that Grosseteste endows the Creation story with a physical interpretation that begins with the creation of light and the identification of light with God, highlighting the thirteenth-century need to buttress belief with rational scientific arguments. Two rational institutions demanded such arguments: the Inquisition and the university.

Colin M. Turbayne, in his article "Grosseteste and an Ancient Optical Principle," notes that "a revolution in the history of science occurred in the thirteenth century at Oxford. The man who started it was Robert Grosseteste." He quotes Dr. A. C. Crombie, who claimed that 'Grosseteste appears to have been the first medieval writer to recognize and deal with experimental verification in science.' According to Turbayne, "The characteristic science in which he applied his ideas on method was optics."[11] Grosseteste's disciple, Roger Bacon, extended the optic research and published three books: *De Multiplication Specierum*, *Perspectiva*, and *Opus Majus*. Witelo published his *Perspectiva*. John Pecham published his *Perspectiva Communis.*[12] Light was the most popular scientific topic in the thirteenth century.

The Islamic source of the thirteenth-century European discussion of light

In the year 1030, Muslim scientist Ibn-Al-Hytham published a revolutionary seven-volume study of optics called *Kitab al-Manazir* or *The Book of Optics*. The first three volumes are devoted to a close analysis of visual perception, while the next three offer a close analysis of reflection, and the last is about refraction. In this work, Al-Hytham challenged the central supposition of Ptolemaic optics that had prevailed in the scientific world for nearly nine centuries. He utilized new terms, theories, and tools that stimulated European intellectuals who adopted and developed them.

Kitab al-Manazir was translated from Arabic into Latin around 1200. The translated book aroused significant interest among the European natural philosophers Robert Grosseteste, Roger Bacon, Witelo, John Pecham, and William of Saint-Cloud, who were influenced by its innovations. The European scholars wrote many studies that followed Al-Hytham's innovative principles, broadening the discourse on light during the thirteenth century.[13]

Testimony exists to the effect that Bonaventure based his theory of the special substantive nature of light on a combination of the physics of light of the Oxford researchers and his own metaphysics of light. Two discourses from two different disciplines – metaphysics and natural philosophy – came together in Bonaventure's thought and made it consonant with its time.

The third source: The monetary revolution – the economic system shifts to the use of money and thinking in terms of money

> Why should not silver be well thought of? After all, with it one buys ermine clothes, abbacies and church benefices, cities and castles, great lands and pretty women. It is silver that disinherits the orphan, absolves the excommunicate, renders justice to the scoundrel and pardons injuries more effectively than pretty words . . . it turns a peasant into a gentleman, makes a happy man of a melancholic, a wise man out of a fool . . . it is silver that ends wars, leads armies, makes ignoble families illustrious. In short, it commands the whole world.[14]

This satire on the influence of money and its nearly all-powerful status in the world, written in the thirteenth century, illustrates the change that occurred in western, central, and southern Europe over the course of this "long century," which lasted from 1160 until the 1330s, when money again became central in the economy and society.

During this "long century," the commercial revolution took place, giving commerce much more centrality in the hitherto dormant economic life of the continent, which had been based almost exclusively on agriculture and a small amount of artisanship. Between the eighth and thirteenth centuries, there was only meagre commercial activity in Europe, and most of it was based upon barter, with almost no involvement of money. During the thirteenth century, commerce returned to Europe and flourished in urban markets and bustling international fairs, among them the particularly prominent fairs of Champagne, Saint Denis, Cologne, and Salonika.

Monetary commerce is based not on barter, but rather on an exchange of goods for money and money for goods. Thus, the commercial revolution of the thirteenth century was also a monetary revolution. Commerce cannot flourish unless there is a mediating measure – some sort of money – that will easily and clearly reflect the values of the goods to be exchanged for it.

Monetary-commercial change resulted in an increase in the quantity of money in circulation, an expansion of the variety of uses of money, and a change in the attitude toward money,[15] to the extent that the society of the time is often referred to as "monetized society."[16] The range of uses of money – from tiny purchases in the market to sales of estates – were all made possible due to the sequent scale of coins that enabled a comparison of values at a high resolution.

The increased amount of money and the creation of the coins' continuum brought about its expanded use in a broad range of economic activity in which it had previously not played a role or in which its presence had been very limited. Monetary payments in villages replaced rents, which had formerly been paid by farmers to landowners in the form of products. Money also took over in paying the corvée, the rent farmers had paid to the nobility in the form of labour on the land. The initiative shown by landowners in demanding monetary rent became a part of the revolution in the attitude toward money. Lands, fields, and forests, previously considered to be a source of goods, were now perceived as a source of money. This change in perception caused the owners of these resources to begin investing money in them in order to improve and restore them so they would yield even more money.[17]

In this way, capital investments in agriculture became widespread, and it became a source of growth, efficiency, productivity, and wealth. Agricultural crops became commodities rather than products. Hired labourers and the payment of wages became widespread and government administrators worked year round for salaries.

In addition, money financed the army and efficiently replaced the feudal military service that had been provided to the lord in return for the use of a parcel of land. Hired soldiers began to fight in return for wages in wartime. According to François-Louis Ganshof:

> Military service was sometimes replaced by a money payment known as scutage . . . In England, at a quite early date, the monarchy permitted money payments in lieu of *servitium militare*, at least in the lower ranks in the feudal hierarchy, and the Plantagents, in the second half of the twelfth century, systematically favoured the substitution of scutage for service, even in the case of tenants-in-chief of the crown . . . The revenue provided by this levying of scutage made possible the hiring of paid troops, who were more trustworthy and more flexible than an army of feudal knights could be at this time.[18]

Money changed the military system that had constituted the feudal regime and gradually undermined it. Money made the troops more trustworthy than the feudal system had through contract, homage, fealty, and kiss (the components that created a vassal relationship during the feudal regime). The new money also generated increased fidelity. Some estimate that in the thirteenth century, the wages for hired soldiers quadrupled the amount of money in circulation.[19]

The development of new commercial techniques contributed to the increase in the quantity of money in the form of bills of exchange.

Originally a provisional substitute for money based on mutual trust among merchants, bills of exchange began to function as money. And so, as Peter Spufford notes:

> the net quantity of silver or gold transported from Bruges to London, or Paris to Florence, did not diminish as a result of the development of bills of exchange, but the amount of business that it represented was increased out of all proportion. The bills of exchange enormously multiplied the supply of money available for international transactions between these cities.[20]

The greater quantity and availability of money enabled increasing numbers of peasants to pay taxes with cash, which replaced tax payments in kind. Hence, it became easier for the kings' courts and their administrators to remain static for long periods, since they collected the taxes in the form of portable money and were no longer dependent on the nearby local crops. In this way, capital cities were founded, since the ruler and his entourage settled in one place and were no longer required to wander from one fief to another. The landlords experienced a similar liberation from the necessity of living in the countryside.[21] In turn, the development of cities populated by the upper classes attracted merchants and artisans who supplied the demands of those classes for many products and luxury goods, which were sold for cash. Money spread through all social classes.

The rapidly increasing use of money had ever-expanding social, economic, and intellectual consequences. As the process of monetization gathered speed, habits of thought and perception initially restricted to those actively engaged in commerce came to be adopted by members of all segments of society. Joel Kaye notes that "among the most characteristic of these new habits were: the focus on monetary profit and loss in a wide range of decision making; the recognition of the importance of detailed written records for this calculation; the resulting broad development of literacy and numeracy; and the translation of qualitative values into quantitative, often monetary, terms as a way to simplify the process of calculation."[22]

St. Bonaventure Lived in the Worldly Monetary World

The following quotation from Diana Wood, scholar of medieval economic thought, addresses Bonaventure's involvement with the notions of commercial life:

> Bonaventure likened God to a trader who had given his son in exchange for the redemption of sinful man. This prompted the question of whether

> God had loved mankind more than Christ, on the basis that if someone gives one thing for another, he values what he receives more than what he gives. Bonaventure did make two crucial points about exchange: that what mattered was subjective value, and that without the hope of gain no one would exchange.[23]

Wood sees this metaphor of God as trader a "bizarre example," yet it reveals how Bonaventure reduced universal relationships to commercial "give and take" relations. For him, God was the ultimate merchant.

The context of the above-mentioned metaphor is a discussion included in Bonaventure's main philosophical book *Commentaria in Quatuor Libros Sententiarum*, a commentary on Petrus Lombardus's basic theological work *Sentences*. Bonaventure wrote it between 1250 and 1252, seven years before publishing his most famous tract, *The Journey of the Mind to God*.

The discussion addresses commerce, merchants, and the core of commercial relations, and shows clearly that Bonaventure had a profound understanding of these. Odd Langholm, scholar of scholastic economic thought, translated some paragraphs from the *Commentaria in Quatuor Libros Sententiarum* that support this evaluation. Here is an example: "It is a foolish merchant who buys a thing more dearly than it is worth, it is to be said that this is untrue except in the case of that merchant who craves the price and intends by trade to enrich himself rather than relieve another's want."[24] The last phrase exposes Bonaventure's attitude regarding the point of commerce being to relieve the wants of others. Hence, commerce is positive and useful. The conclusion of this discourse is that "God does not trade in order to increase his own utility but in order to relieve our want."[25] God, then, is a supremely benevolent merchant, one who completely realizes the ultimate telos of commerce.

Bonaventure studied at the University of Paris beginning in 1235. He received his Master of Arts degree in 1243, the same year in which he joined the Franciscan Order. His name was then changed from Giovanni di Fidenza to Bonaventure. He was nominated master of the Franciscan school in Paris in 1253. He joined the University of Paris as a master in the Faculty of Theology alongside the Dominican Thomas de Aquinas at the end of 1256, and was nominated head of the Franciscan Order in 1257. Hence, he lived in Paris for twenty-two years as a secular student or master, and during this long period became deeply acquainted with the city, which was the most economically active in Europe at the time. This acquaintance developed into a deep and extensive understanding. Regarding the role of Paris as an economic centre, Virginia Wylie Egbert notes:

> Any visitor to mediaeval Paris would have soon discovered that the peak of its activity centered on the two bridges connecting the island of the cite with the Right and Left banks . . . and the Grand Pont, spanning the northern branch of the Seine, led to the Right Bank . . . In the twelfth century, philosophers known as Parvi-Pontins lived on the Petit Ponts; later they were succeeded by merchants. Here tolls were collected on food supplies and merchandise brought into the city by land. This bridge became such a hustling center of trade that Joinville, in describing the burning of the bazaar at Damietta by Turks in 1249 wrote, 'The damage that followed from this was as great as if someone were, tomorrow, to set fire to Petit-Pont in Paris'.[26]

Paris's merchants comprised its central economic and political sector during the flourishing first half of the thirteenth century. "A particularly important group were the money-changers. These were Italians for the most part . . . They were obliged to pursue their activities out in the open on the Grand Pont . . . henceforth known as the *Pont des Changeurs* (Exchange Bridge)."[27]

Money was also an integral part of Parisian students' life and culture as it was reconstructed by the medievalist Charles Homer Haskins, who used the preaching literature as his main historical source, since "the thirteenth century was an age of preaching . . .[and] the enormous mass of sermons which has come to us from that period is well worthy of examination for the light it throws upon the University of Paris and its life."[28] The students were served by a variety of peddlers, butlers, and assistants, who kept good seats for them in lecture-halls, carried their books to class, and supplied them with food and sweets. All of these services were paid for by the wealthy students.[29] Thus, as a student, Giovanni di Fidenza (who would later become Bonaventure) was surrounded by money and exposed to the monetary culture. He was acquainted with the reality in which coins paid for everything from personal services and small items to real estate. The presence of money in Parisian life was ubiquitous, and it was a universal measure.

The fourth source: The passion for measuring

A fourth change brought about by the monetary revolution was an overwhelming urge to measure that gripped thirteenth and fourteenth century scholars. The passion for measuring and quantifying spread into every sphere of thought, including theology. The aspiration to create a quantified hierarchy was applied to many spheres and created a desire to quantify qualities.

Scholars of the period discovered that money was a universal tool for measuring and comparing products throughout the world, even those that were completely unrelated to one another. Even nature could be measured if only the means were found. Not only nature was measurable. So were human characteristics, including such attitudinal qualities as professionalism, humility, concern, devotion, and more.

According to Joel Kaye, the development of a more complicated concept of money was what generated this passion for measuring. His thesis is the main inspiration for my studies. This money concept, which Kaye called "administrative money" can be seen as revolutionary. Kaye explains:

> Where money in exchange serves as a medium for connecting diverse goods and services by making them commensurable, money in administration serves as a medium for dividing values along a single continuous axis. In exchange, the primary function of money is to facilitate the finding of an equalizing point or range (latitude). In administration, equality remains the ideal, but the goal is to determine points (or latitude) or proportionally equal divisions along a continuum of value.[30]

This is administrative money because, as Kaye elaborates:

> money measurement found its most rational application within the growing administrative bureaucracies of the century – city governments, royal governments, the administration of diocese, chapter, and order, ballooning bureaucracy of the papacy at Avignon, the expanding university itself. Bureaucracies require standardization to function. The first step in bureaucratic standardization is the introduction of money-price as standardized measure.[31]

By the thirteenth century, money served as the prime bureaucratic instrument in the quantification of qualitative social merits and responsibilities. Hence, the bureaucratic quantification of social qualities through the use of a monetary scale was established in society and paved the way to the strong drive to measure and quantify, even in philosophy and theology.

The sequence of the money chain from the coin of the lowest value to the most precious one enabled people to quantify everything and construct a continuum from the cheapest to the dearest. This scale was conceptually transferred from the material world to the spiritual domain. This analysis is in line with Marx's well-known insight that men's ideas are the most direct emanation of their material existence.

The ultimate connection between money as a theological measure and the Franciscan order - the indulgences

The Regula Bollata, issued by Franciscus of Assisi, the founder of the Franciscan Order, in 1223, prohibited the friars from possessing any property. In 1250, this regulation gave rise to a long debate, performed as a play in three acts, between the Franciscans and the secular masters of the University of Paris. Although the main controversy between the two camps concerned the definition of property, it included another question regarding the definition of the boundary between money and non-money. This debate compelled the Franciscan Order, headed by Bonaventure, to organize its 255 directives into a coherent codex that discussed the strict limits of using money.

In spite of the strict prohibition on possessing property or money, the Franciscan Order was deeply involved in one of the most prominent barter exchanges, which was initiated in the thirtieth year of the thirteenth century – money for redemption of the crusade vow. Christoph T. Maier explains:

> The crusade vow differed from most other religious vows in two principal ways. Firstly, it was accompanied by the promise of a plenary indulgence, and, secondly, its execution was not merely an individual act of devotion and penitence, but it was also part of a collective effort in the context of the *negotium crusis* [crusade business]. The redemption of a crusade vow was technically a form of vow commutation substituting an adequate penitential exercise *in lieu* of crusading, in this case the payment of a certain amount of money in aid of the crusade.[32]

But how much? The common rate was the amount of money that financed a common crusader. A poor man was allowed to finance less than the usual sum. These calculations were too vague and led to the development of more accurate criteria, all of which converted quantifications and qualifications into money. The most active redemption collectors were the Franciscan friars, who had been ordered to refrain entirely from dealing with money. In this way, money paved the way to paradise, since it purified those who paid. In the thirteenth century, money became the omnipotent cleanser of sin.

A prologue as an epilogue: The first step in *The Journey of the Mind to God* is the light

"Prologue: as I begin, I call upon the First Beginning. The father of Lights from whom all illumination descends, from whom every good gift and every perfect gift derives . . . Give light to the eyes of our mind . . ."

God is light that directs mankind to ascend the ladder of light in order to be closer to the absolute God/light. *Itinerarium Mentis in Deum* by Bonaventure is the directory, the guide. It is built on six steps, of which the first is the identification the highest (God) to the lowest (Earth). The ladder of money became a ladder of light, with God the absolute light: "through a similitude of the divine light that shines upon us, and in that light itself."[33] Merging with God was the uppermost stage. Jacob's ladder, which connected earth to heaven, became, in Bonaventure's theology, a ladder of light whose continuum was the same as the sequence of coins, but from the highest (God) to the lowest (Earth). The ladder of money became a ladder of light.

Notes

[1] Frederick Copleston, *A History of Philosophy*, vol. 2., *Mediaeval Philosophy: Augustine to Scotus* (London: Burns and Oats, 1966), 274.
[2] John F. Quinn, *The Historical Constitution of St. Bonaventure's Philosophy* (Toronto: Pontifical Institute of Medieval Studies, 1973), 239.
[3] Bonaventure, *The Journey of the Mind to God,* ed. Stephen F. Brown, trans. Philotheus Boehner (Indianapolis and Cambridge: Hacket Publishing Company, 1993), 2.
[4] Quinn, *The Historical Constitution*, 512.
[5] Robert S. Lopez, *The Birth of Europe* (New York: M. Evans and Company, 1967), 366.
[6] Ibid.
[7] Robert Grosseteste, *De Luce* (*On Light or The Beginning of Forms*), trans. Clare C. Riedl (Milwaukee: Marquette University Press, 1942), 10.
[8] Ibid.
[9] Ibid.
[10] Ibid., 1–2.
[11] Colin M. Turbayne, "Grosseteste and an Ancient Optical Principle," *Isis* 50, no. 4 (1959): 467.
[12] Mark Smith, "What Is the History of Medieval Optics Really About?," *Proceedings of the American Philosophical Society* 148, no. 2 (2004): 182–83.
[13] Ibid.

---

[14] Joel Kaye, *Economy and Nature in the Fourteenth Century: Money, Market Exchange, and the Emergence of Scientific Thought* (Cambridge: Cambridge University Press, 1998), 18.
[15] Peter Spufford, *Money and Its Use in Medieval Europe* (Cambridge: Cambridge University Press, 1988), 240.
[16] Kaye, *Economy and Nature in the Fourteenth Century*, 170.
[17] Spufford, *Money and Its Use in Medieval Europe*, 245.
[18] François-Louis Ganshof, *Feudalism* (London: Longman, 1970), 90.
[19] Diana Wood, *Medieval Economic Thought* (Cambridge: Cambridge University Press, 2002), 79.
[20] Peter Spufford, *Handbook of Medieval Exchange* (London: Offices of the Royal Historical Society, 1986), 33-34.
[21] Ibid., 249.
[22] Kaye, *Economy and Nature in the Fourteenth Century*, 16.
[23] Wood, *Medieval Economic Thought*, 150–15.
[24] Odd Langholm, *Economics in the Medieval Schools: Wealth, Exchange, Value, Money and Usury according to the Paris Theological Tradition, 1200–1350* (Leiden: Brill, 1992), 157.
[25] Ibid.
[26] Virginia Wylie Egbert, *On the Bridges of Medieval Paris: A Record of Early Fourteenth-Century Life* (Princeton: Princeton University Press, 1974), 21–22.
[27] Colin Jones, *Paris: Biography of a City* (London: Penguin Books, 2006), 54.
[28] Charles Homer Haskins, *Studies in Medieval Culture* (Frederick Ungar Publishing: New York, 1929), 38.
[29] Ibid., 66–67.
[30] Kaye, *Economy and Nature in the Fourteenth Century*, 174.
[31] Ibid., 169.
[32] Christoph T. Maier, *Preaching the Crusades: Mendicant Friars and the Cross in the Thirteenth Century* (Cambridge: Cambridge University Press, 1994), 135.
[33] Ibid., 37.

CHAPTER NINE

"THE CUTTING OF LIGHT THROUGH THE CONTINUUM OF TIME":[1] LIGHT IN SHIMON ADAF'S *THE WEDDING GIFTS*

RINA JEAN BAROUKH

> As a photon wends its way through an arrangement of glass panes and mirrors, its path remains ambiguous. It essentially takes every possible path available to it. This ambiguity remains until observation by a conscious observer forces the particle to decide which path it had taken. Then the uncertainty is resolved – retroactively – and it is as if the selected path had been taken all along.[2]

Part one

The Wedding Gifts, by Shimon Adaf, published in 2014, is a novel that consists of several parts, narratives, and plots. The parts are diverse in style and genre and include poetry, a realistic novella, and a cyber-punk story. The novel is complex and has many diverse themes. One of the most important and interesting themes of the novel that runs through all of its parts is light. Furthermore, the movement of light, in the form of rays, can offer an explanation of the way in which the different parts of the novel are connected and help to explain the reading process of this unique and complex work. In this chapter, I will show why light is so important for the understanding of *The Wedding Gifts* and examine the different ways in which light connects to its major themes. Since the novel is highly complex, I will be able to address only a small number of its themes here. As I will show, *The Wedding Gifts* is full of light, and, naturally, of the darkness that surrounds it.[3] I claim first that light is one of the major themes in this novel, and second, that the way in which light moves and operates is what connects the separate parts of the novel.

Shimon Adaf, born in 1972, writes in almost every genre: poetry, fiction, science fiction, detective fiction, and more. He tends to combine different genres and use them in innovative ways. Adaf has published twelve books so far, three of poetry, eight of fiction, and a non-fiction book, co-authored with Lavie Tidhar, published in 2016.[4] He has won several prizes for his writing, including the 2012 Sapir Prize for Literature for his novel *Mox Nox.*

The Wedding Gifts, with its seven parts, some of them written in verse and some in prose, may seem at first an arbitrary collection of disconnected and fragmented stories. I would like to suggest that, on the contrary, the novel has a precise and remarkable structure. This structure frames the disjointed narratives together with the leitmotif of the novel. Of the novel's seven parts, three of them – the first, fifth, and seventh – are written in poetic verse, while the remaining four are written in prose. Of these four parts, three have epigraphs drawn from various Hebrew texts. All of the epigraphs also have themes connected to light.[5] I would like to suggest that each epigraph comprises not only an introduction to each part, as expected, but can also be regarded as a separate entity in its own right, bringing the total parts of the novel to ten.[6] Given that a number of the novel's themes concern redemption, messianic motifs, and different aspects of light, its division into ten parts brings to mind the Kabbalistic structure of the notion of ten *sfirot*, *ẓimẓum*, the breaking of the vessels, and *tikun*.[7] To briefly summarize this complex narrative: once God (as light) filled the entire world, but in order to make room for Creation, he had to remove himself from the world (*ẓimẓum*). His light poured into vessels that were too weak to hold it, and they broke. But light was trapped inside these broken vessels (our world being one of them), in the form of *niẓoẓot*, or sparks of light. The sparks of light, trapped and isolated, correspond with the different parts of Adaf's novel, each one being a separate story that holds the light within it, but is separate from the other stories. As I will demonstrate, light will also bring these separate parts together.

In any case, the novel's division into parts is full of significance, since the numbers of these parts are all typological, numbers that hold a special meaning in the world of Hebrew thought: three epigraphs, seven parts, and the sum of ten, as I mention above, relate to the *sfirot*. As described by Rachel Elior, the number seven is also connected to light, since it goes back to the seven days of Creation, which are commemorated in the light of the seven candles, one for each day of the week, that were lit in the *mishkan* (Tabernacle) and the seven-lamp menorah in the Temple.[8]

Beyond the obvious stylistic differences between poetry and prose described above, the parts of the novel that are written in prose differ from one another in style and genre and are concerned with different times and places. Thus, the connection between the parts is not immediately apparent. Although some of the characters appear in more than one of the different parts of the book, in order to discuss the story as a whole, I will attempt to reconstruct it here.[9]

As mentioned above, the novel opens with a poem. Following the poem, a seemingly realistic novella appears, telling the story of Salior Fadida and his family, especially his son, Noam, who disappears at the end of the narrative segment (upon which I will elaborate below). In the third part, we have the story of a writer who spends the summer at an artists' colony in Johnson, Vermont by a lake called Eden.[10] Once there, he has a series of meetings with other artists, meets a robotic oracle,[11] and undergoes a techno-mystical experience. The fourth part of the book is a futuristic, science fiction (or cyber-punk) story whose action takes place on a planet called Aridu.[12] This part combines the story of Mrs. De Leon, a teacher from the high school that Noam (the same Noam from the second part) attends. We learn that after his disappearance, Noam reappeared on Aridu, suffering from memory loss and not knowing exactly how he got there. Mrs. De Leon tries to help him find out what happened to him, and makes discoveries about herself as well. In particular, she learns about the son, Soon, she once had, whose memory she has lost. The fifth part is a cycle of poems, while the sixth part takes place in the futuristic world of Mrs. De Leon, which goes back in time to the story of her and her son. Finally, as mentioned, the novel ends with a poem.

The various connections between these segments are not apparent from a casual reading of the text. In addition, since all the stories begin *in media res*, the moves from one to another take place without warning or explanation. Repeatedly, readers are required to discover where and when they are, and speculate on how exactly they got there, which is an important effect of the text. It seems at times that each part of the book is a world of its own, a planet, perhaps, surrounded by darkness through which light still manages to shine, and it is this light that is the subject of this chapter.

Part two

The second part of the novel (following a poem) is a realistic novella about Salior Fadida and his family, who live somewhere in southern Israel. Salior goes back to live with his mother after separating from Yulia, his wife. Yulia lives not far away, and their three children, Noam, Shira, and Shalom go back and forth between them. Salior is about thirty years old, but seems much older. He is overweight, has breathing, sleeping, and drinking problems and struggles to make a living. He installs security cameras for a living. One day, when he installs cameras at the supermarket, he gets to know Xenia, a nineteen-year-old girl who works there, and a relationship develops between them.

But beneath the concrete daily life that the novella describes in realistic detail, an abyss opens, and Noam, Salior's oldest son, only six and a half years old, disappears into it.[13] Noam's disappearance is not described, and is revealed only in the last lines of the novella, when, with his disappearance, the story ends. The last sentence of the novella is Yulia's question to Salior: "Where is Noam? Tell me he's with you. I can't find him anywhere."[14] With this unanswered question, the realistic mode of description and storytelling is exposed as a deception and the realistic appearance of the novella is shown to be misleading.

This revelation is, of course, part of the literary act itself. The story ends abruptly with Noam's disappearance, with no explanations, closure, or epilogue. The text does not reveal what happens afterward. Here and throughout the novel, the narrative moves from one story to another, so the reader is forced to move forward, somewhat like the speaker of the poem mentioned above, with which the novel begins, with no explanations.

Part three

Light, or *ḥashmal*, electricity, in Hebrew, appears in many different ways in each part of the novel. The word *ḥashmal* is taken from the prophecy of Ezekiel: "And I looked, and, behold, a whirlwind came out of the north, a great cloud, and a fire infolding itself, and a brightness was about it, and out of the midst thereof as the colour of amber, out of the midst of the fire."[15]

No one knows exactly what Ezekiel meant by the word *ḥashmal*, but it has always been linked with luminosity.[16] The word was translated as "amber," or electron, in the translation of the book of Ezekiel, around the second century BCE in Alexandria, as part of the Septuagint.[17] In Modern Hebrew, the word for electricity is *ḥashmal*, serving as a reminder, though slight, of that mystic luminosity of Ezekiel's vision. The word holds both meanings – mundane, modern, manmade technological electricity, and the transcendent luminosity that is also the light of God. This duality of light is emphasized in every part of *The Wedding Gifts*, so light can be seen as full of meaning or as having no meaning at all.

In the novella in which Salior is the protagonist, light is manifested in many ways, in candles, flashlights, and digital and electronic devices that flicker and flash in the darkness. Natural light is also present in the form of lightning, sunshine, and moonlight, which are described in many scenes. It is possible to make a categorical distinction between artificial and natural light in the novella and in the novel as a whole, but the crucial point is to regard both types of light as being both transcendental and mundane, to hold both possibilities at the same time, like the word *ḥashmal*. For example: "a faint field of light fell on the leather book spines of the *Chumash*, on the letters written in gold"[18] or "a thin glow washed the furniture when the light of the phone floated before [Salior]."[19] In these and many other examples in this part of the novel, light from all origins, natural and man-made, has special capacities. Therefore, depending on interpretation, transcendental characteristics can be given to artificial light and taken away from natural light, making the mythical and the ancient similar to the technological and the innovative.

The combination between the transcendental and the mundane can be seen in the novella in places where the characters are distinguished by their attitude towards light and their relationship with it. Salior, as described before, is, as the narrator puts it, the most "solid" body in the house,[20] and his "heaviness" is mentioned several times, but his name offers a different perspective. Sa-li-or (שא-לי-אור) literally means "carry-the-light." But although he has an inherent connection to light through his name, he cannot relate to it. Since he is "solid," no light can shine through him. Xenia, the young girl he is interested in, is his opposite. She sees light as the light of God in every aspect of nature, in a pantheistic way. One day, she looks at the sky, and the following conversation ensues:

> You like looking at the sky, he said.
>
> Yes. Who knows, maybe I'll be able to see God eventually.
>
> He laughed.
>
> I mean it, she said, when I was a little girl, my mom promised me that God was watching us from the clouds.
>
> You're not a little girl anymore.
>
> No, but she was right. If you look long enough, there is a detail that doesn't fit, something opens after that, a deeper depth, it's in the trees, and it's in the bugs, and it's in the blowing of the wind. It's easier to see in the clouds. . . .
>
> You actually believe in that crap you said about the trees.
>
> She was filled with distance, something in her eyes withdrew. She looked down. Trembling, she said, in a restrained voice, I'd better go home now.[21]

Salior, the one who is supposed to carry the light, cannot see it. Like a mirror image of the blind Tiresias, who sees the light of truth better than anyone, Salior, the one who has the light, is completely blind to it. For him, it holds no special meaning. But Xenia and Noam, his son, share a similar connection to nature and to light. For them, it holds a mystery, a transcendence. Here we see that the question is not about what kind of light it is, whether it is the biblical vision of *ḥashmal*, or everyday electricity. The question is rather, are you able to see it and what meaning does it hold for you?

Why is Salior unable to see the light that he carries? It is suggested that his love for Yulia has blinded him. "[L]ove shuts your eyes," Salior says.[22] So by marrying Yulia (or Juliet, who, as we know, Romeo compared to the sun), he loses his own light, only reflecting back the light around him, like a moon. Perhaps his choice to follow physical passion means that he has to give up a more spiritual existence: "I followed the heart,"[23] Salior tells Xenia, describing to her how he left everything when he was sixteen to follow Yulia. Salior also remembers what his friend, Hamurabi, told him when they were both sixteen. Hamurabi said that Salior was "Thinking through [his] dick. . . . and that from the first moment [Hamurabi] saw Yulia, he knew she would be [Saliors'] disaster."[24] This is one possible meaning and I will offer another explanation below.

Salior's unawareness of the light and his lack of will to give it any meaning create a barrier between him and the ones he loves. Noam attempts to share a few of his revelations with his father, peculiar things about time and space that can retroactively explain his disappearance, but Salior rejects every attempt to see beyond concrete reality. Since he cannot provide rational explanations for Noam's revelations, he ignores them, and ultimately loses him.

Unlike Salior, and like Xenia, Noam shows a strong connection to light. He, however, like other children in this novel, has a calling, and represents the possibility of redemption, yet only to a certain degree, since ultimately, this possibility remains unfulfilled. In Adaf's novels, children on the verge of adolescence tend to have the strongest connection to the world and to whatever powers control it.[25] However, the narration makes clear that they both see things that people around them do not see. Noam realizes that time does not move in straight lines. He finds a point in Salior's old room "where he puts his watch and it stops"[26] that he calls a "zero point."[27] Indeed, one night, during a lightning storm, something strange happens. As Noam looks out the window, his portrait is etched by lightning onto the surface of the glass: "On the glass surface, a pale portrait of Noam was burned, a ghostly image was looking outside the window in wonder."[28] Salior's old room becomes a *camera obscura*, freezing a moment in time. This strange photograph takes on a different meaning for each character. For Salior it is only a natural, if rare, phenomenon, while for his religious family members it is a miracle that needs to be brought to the rabbi. For Xenia it is a work of art created by nature. Yet for Noam, who will disappear soon, it is like a long-awaited meeting or a call. The lightning here combines the electricity of the *ḥashmal* and the luminosity of light. It is both a natural phenomenon, but the portrait it creates suggests a supernatural meaning. This portrait is the only thing his family will have of him after his disappearance to a different time, a different place. As Erich Neumann puts it: "in every place where a mystic experience occurs, the numinous layer bursts throughout the fixed stability of any given universe. But crossing this border entails . . . detachment and a tear from humanity and from the worldly reality into the in-human, into outer-world spaces."[29] Similarly, in Adaf's novel, the supernatural phenomenon creates a moment in the story that transforms it from a realistic novella to a fantastic one. Thus, Noam is torn from his family and from the story much in the way that Neumann describes the tearing from humanity that takes place when a mystical experience occurs.

Light in this novel is both an eternal flow, like the *ḥashmal* that flows through time into our reality, and the ability to freeze a particular moment, like Noam's portrait. In the novel, light captures time. In several instances, as with Noam's portrait, light operates as a camera, freezing a moment in time. Time is then like a shard upon which the events are etched in light: "A brightness burst into the space, burning both of them [Salior and the policeman, a minor character in the novella] onto the plate of the passing second. . . . the infinite brightness shone, and that blink of an eye was captured, kidnapped from the realm of its being."[30] Salior's security cameras exemplify this dichotomy between time as one continuous movement on the one hand, yet made from a series of single, frozen images, on the other. Light can represent a continuum, from Ezekiel's visions to our reality, but it can also freeze and isolate a fragment in time. This is the double meaning of the quote: "the cutting of light through the continuum of time."[31] Time flows and moves, but it is also a series of single, frozen images. Similarly, the novel may also be read as a continuum, even though it is made up of separate parts.

Part four

I have mentioned the fall, or falling movement that appears in Salior's story. This movement accompanies this story as well as other parts of the novel and is also connected to light. First, the novel begins with a falling movement in the poem that effectively serves as its prologue. It describes a meeting that transports the speaker from an existence out of time and space into the world: “I met the world / in an outer point.”[32] He enters a world of “Airports, rails / steel and electricity,” and is violently and aggressively pushed into a world of “sex!” and “great cities.”[33] Facing these temptations, this world full of possibilities, the speaker utters a single word of resistance: “No,”[34] which he repeats in the phrase “I said / no” again and again. The speaker enters this world by force, not by choice. The verbal resistance of the speaker is futile, since the world does not need his consent, but only himself, his existence, his body. The world will do what it likes with him. The poem ends with the lines:

In an outer point before
  Axis I said
  No
  No
  Let me go
  And the world said, will you look at that
  In the middle of speech how much
  Of you I already have, how much
  Of you is already here.[35]

The poem describes the descent of the speaker that begins outside of time and space, and ends “here” in the physical world. Like one who falls from heaven, like an angel falling from the sky, “here” is where the body of the speaker now lies.

This falling movement relates to another story that connects Salior, Noam, and light. It is a story of a fallen star, or angel, the two-faceted story of Lucifer and Heilel Ben-Shahar:

> How art thou fallen from heaven, O Lucifer, son of the morning!
> How art thou cut down to the ground, which didst weaken the nations!
> For thou hast said in thine heart, I will ascend into heaven, I will exalt my throne
> above the stars of God: I will sit also upon the mount of the congregation,
> in the sides of the north:
> I will ascend above the heights of the clouds; I will be like the most High.
> Yet thou shalt be brought down to hell, to the sides of the pit.[36]

Heilel Ben-Shahar is the Hebrew for morning star, or the planet Venus, which also appears in disguise as the planet Zimbir later in the novel. In the vulgate translation, or mistranslation, it became Lucifer, or Satan, the fallen angel, whose name means "the light-bearer" (from *lux*, or light, in Latin). If we combine the stories of Heilel Ben-Shahar and Lucifer, we get a double image of a shining star and a fallen angel, both of whom hold a certain luminosity (fig. 9.1). This double image of the star and the angel is reminiscent of Salior, who, as mentioned, carries the light, but denies it. Like Lucifer, he is someone who had the light, but lost it. The image is inherently paradoxical, since both the star and the fallen angel carry the light, lead it, and are close to it, but it is no longer theirs. Connecting Salior, Heilel Ben-Shahar, and Lucifer explains why Salior resists light and denies any sign of spirituality. Salior, who carries the light, but is blind to it, acts throughout the novel as if he has been betrayed, as if he has been given a promise that was not fulfilled. But unlike Lucifer, who was cast down to hell, Salior falls, as demonstrated, into his physical body. So Salior's fall, in effect, encompasses two falls: the fall from paradise of Adam and Eve, and the fall of Lucifer. Lucifer's fall is also reflected in other fallen figures mentioned in the third part of the novel, such as the poets Arthur Rimbaud and William Blake,[37] who also wrote of the fall from heaven, which correlates to the falling theme of the novel.[38] The falling movement also resonates in Noam's fall to another world, another time, and another story.

Essentially, the novel suggests that being chosen, for whatever reason, means being torn from the fabric of your life, much as the characters are torn from the text of this novel, time and time again, like Noam, who disappears in the novella and reappears in a different story. There is a high price to pay for being close to the heart of light, to *ḥashmal*, to the secret of things. It means being left alone outside of the flow of time, as Neumann suggests. In this novel, Adaf examines the price of being chosen. He exposes the violence behind the promise, the betrayal that is already enfolded in the moment of anointment. It is the story of all who "fall without / the possibility to fly.[39] You can hear it in the call of Jesus on the cross, "My God, my God, why have you forsaken me?,"[40] which is the similar to what Salior's brother says after their father dies: "I loved him and he destroyed me,"[41] which in turn echoes T.S. Eliot's Gerontion, who says, "I that was near your heart, was removed therefrom."[42] Every section of this novel embodies this fall and retells the eternal myth of being chosen and punished for it at the same time. "The world does need a Messiah," the narrator says later in the novel, "but it is not a job for a hero, it never was. What we need is a villain."[43] The repeated fall is combined in

Fig. 9.1: Gustave Doré, *Me miserable! Which way shall I fly, Infinite wrath, and infinite despair?* Engraving, illustrated for Milton's *Paradise lost*, Image courtesy of the Poetry Collection of the University Libraries, University at Buffalo, The State University of New York.

this novel with another fall, the fall from paradise that has trapped humankind in a body that needs and wants. So when the characters fall once, they fall twice – once far from what they know and love, and once into the body from which they cannot escape. As the speaker puts it in the last poem of the novel:

> I'm hungry
> Lord
> why hast thou cursed me twice
> once with the body
> once with the body
> without which I can
> never know
> you.[44]

With these words, the poem and the novel end. What started "[i]n an outer point before Axis" ends with the hungry body, longing to know that it is not alone in the darkness into which it was pushed by force, not by choice. This fall, I suggest, is not only a theme in this novel, but also a way of understanding the way it should be read, and the way in which the reader must process it from beginning to end. Each section tells the story of a different fall, but the novel as a whole is one continuous falling movement, creating an Axis Mundi, from beginning to end. Since the fall in the novel is combined with the falling light, that of Lucifer and Heilel Ben-Shahar, the light and the fall *become* the structure of the novel, as well as a major theme in it. In this way, each story or section holds its own light within it, but the whole novel is like a single ray of light that incorporates all these worlds, falls through all these stories, like "the cutting of light through the continuum of time."[45]

Notes

[1] Shimon Adaf, *The Wedding Gifts* (Or Yehuda: Kinneret, Zmora-Bitan, Dvir, 2014), 47. All translations from the novel are mine unless otherwise noted.

[2] Ray Kurzweil, "A Note for the Reader" in *The Age of Spiritual Machines: When Computers Exceed Human Intelligence* (London: Penguin Books, 2000), v.

[3] The contrast between light and darkness can be seen in the graphic, black and white cover image of the book as well as in its illustrations.

[4] See my review: Rina Jean Baroukh, "'The Speculative is the Actual': On Art and War: Poetry, Pulp and Politics in Israeli Literature," *Iton 77*, 389 (2016): 16–17 (Hebrew).

[5] The epigraphs are taken from the Babylonian Talmud, the *Book of Palaces*, and the *Book of Visions* by Hayyim ben Joseph Vital.

[6] This assumption is strengthened by the fact that each epigraph is printed on a separate page, before the part it is attached to, with an illustration of its own.

[7] There are many books about the *sfirot*. For a short introduction on this subject, see: Gershom Scholem, *On the Kabbalah and Its Symbolism,* trans. Ralph Manheim (New York: Schocken [1969] 1996), 109–17.

[8] Rachel Elior, "The Cycles of Divine Time, the *Merkavah* Tradition and the Dispute over the Calendar," in *Lights: In Literature, Art, and Jewish Thought,* eds. Emily D. Bilski, Amitai Mendelson, and Avigdor Shinan, (Tel Aviv: Am Oved, 2005), 94 (Hebrew).

[9] In an interview Adaf gave while working on *The Wedding Gifts*, he described its structure as follows: "it begins with a poem that gives birth to a realistic novella, that gives birth to a non-fiction piece discussing several themes, that give birth to an alternative universe police novella, that gives birth to a cycle of poems, that gives birth to a space opera kind of novella, that gives birth to religious hymns." Rochita Loenen-Ruiz, "*Author Interview: Shimon Adaf*." https://chieandweng.wordpress.com/2013/10/16/221/. Accessed September 9, 2016.

[10] "Eden" means paradise in Hebrew, and may also refer to "Gan Eden," the Garden of Eden. Since the writer goes back to Eden, he can be one of the persons who "entered in their lives to the garden of paradise." See Avigdor Shinan, "Entered in Their Lives to the Garden of Eden," in *A Garden Eastward in Eden: Traditions of Paradise – Changing Jewish Perspectives and Comparative Dimensions of Culture*, ed. Rachel Elior, (Jerusalem: Magnes, 2010), 225–38 (Hebrew).

[11] The scene in the novel where the narrator meets the AI robot called Bina is based on an actual AI, humanoid robot called BINA48 (Bina stands for: **B**reakthrough **I**ntelligence via **N**eural **A**rchitecture). While this is a real story, the word *bina* also means wisdom in Hebrew. On BINA48 see https://www.lifenaut.com/bina48/. Accessed January 23, 2017.

[12] Aridu is an ancient Sumerian city, and the novel is strongly connected to Sumerian mythology. I will not address this issue here.

[13] Adaf, *The Wedding Gifts,* 84.

[14] Ibid., 133.

[15] Ezek, 1:4.

[16] On the etymology of the word *ḥashmal* and its translations see: Alon Idan, "Electricity." http://www.haaretz.co.il/magazine/the-edge/mehasafa/1.1899187. Accessed September 6, 2016.

[17] Idan, "Electricity."

[18] Adaf, *The Wedding Gifts,* 25.

[19] Ibid., 53.

[20] Ibid., 20.

[21] Ibid., 78.

[22] Ibid., 56.

[23] Ibid., 89.

[24] Ibid., 119.

[25] See Rina Jean Baroukh, "'Your Light Has Come': Fantasy and Reality in Shimon Adaf's *Sunburnt Faces*" (paper presented at the CRSF (Current Research in Speculative Fiction) conference, Liverpool, June 27, 2016).

[26] Adaf, *The Wedding Gifts,* 83.

[27] In mathematics, a "zero-point" is a fixed point of reference for a coordinate system.
[28] Adaf, *The Wedding Gifts,* 96.
[29] Neumann, *Der Mystische Mensch*, 71.
[30] Adaf, *The Wedding Gifts*, 46.
[31] Ibid., 47.
[32] Ibid., 7.
[33] Ibid., 8. The phrase "great cities," or in Hebrew: "ערים גדולות ובצורות", is taken from Deut. 1:28: "Whither shall we go up? our brethren have discouraged our heart, saying, The people is greater and taller than we; the cities are great and walled up to heaven; and moreover we have seen the sons of the Anakims there."
[34] Adaf, *The Wedding Gifts,* 7, 8.
[35] Ibid., 8. There is another translation of this poem, by Edward Levin, in: Dorit Lemberger, *A Red Rose in the Dark: Self-Constitution through the Poetic Language of Zelda, Amichai, Kosman, and Adaf* (Brighton MA: Academic Studies Press, 2016, forthcoming), 130–32. I thank Shimon Adaf for referring me to the existing translations of the poems in the novel.
[36] Isa. 14:12–15.
[37] Adaf, *The Wedding Gifts,* 152. Gershom Scholem comments on these figures: "certain authors, disregarding or rejecting all traditional authority, describe their mystical experience in resolutely secular terms, yet clothe their interpretation of the same experience in traditional images. This is the case with Rimbaud and more consistently with William Blake. They regard themselves as Luciferian heretics, yet their imagination is shot through with traditional images." Scholem, *On the Kabbalah and Its Symbolism,* 16.
[38] There is a possibility that the title of the novel relates to Blake's *The Marriage of Heaven and Hell.*
[39] Adaf, *The Wedding Gifts,* 267.
[40] Mark 15:34.
[41] Adaf, *The Wedding Gifts,* 92, 132.
[42] T. S. Eliot, "Gerontion," *Collected Poems 1909–1962,* (New York: Harcourt Brace and Company, 1963), 31.
[43] Adaf, *The Wedding Gifts*, 259.
[44] Ibid., 268. Translated by Jenny Minniti-Shippey, with Shimon Adaf. I thank Shimon Adaf and Jenny Minniti-Shippey for allowing me to use this translation.
[45] Ibid., 47.

CONTRIBUTORS

Yair Barak's areas of research are economic history, history of economic thought, economy as culture. He is affiliated with The Zvi Yaavez School of historical studies at Tel-Aviv University and the Gildenhorn Institute for Israel Studies and the Economics department at Maryland University, Washington D.C.

Niharika Dinkar teaches art history at Boise State University. She studied at the National Museum in New Delhi and the State University of New York at Stony Brook. Her work has focused on art and politics in nineteenth-century Indian painting and visual culture.

Merav Fima holds an MA in English Literature and Creative Writing from Bar-Ilan University (Israel) and a BA in Western Society and Culture and Art History from Concordia University (Montreal, Quebec, Canada). She previously worked as an editor at The Israel Museum, Jerusalem, and currently serves as the Coordinator of English Publications at Ben-Gurion University of the Negev.

Stephan Gregory studied philosophy and German literature at the Universities of Munich and Vienna. He has been teaching at Merz Academy in Stuttgart and Bauhaus University in Weimar. Since 2014 he has been conducting the research project "Mimesis Tropical. Jesuit Styles of Imitation in Colonial Brazil" within the research team *Media and Mimesis*, based in Weimar.

Sharron Hass is a Poet, essayist, and lecturer. She is a graduate of the Classics Department at Tel-Aviv University, and holds an M.A. degree in Religious Studies. She lectures on literature and poetry at Alma Institute in Tel-Aviv, and at Tel-Aviv Museum of Art, and teaches in the Creative Writing Program at Tel-Aviv University. In 2005, Hass received a Fulbright Scholarship to participate in the International Writers Program (IWP) at the University of Iowa. She has published five books of poems: *The Mountain Mother is Gone* (1997, Hezi Laskley Award and the Art Council Prize), *The Stranger and Everyday-Woman* (2001, the Prime-Minister Prize for Poetry), *Subjects of the Sun* (2006), *Daylight* (2011, The Bialik Prize), *The Music of the Wide Lane* (2015). Translations of her poems into English and other languages appeared in anthologies of contemporary Israeli poetry, such as: *Chacune a un nom: Femmes poetes et artistes d'Israel* (Caracteres, 2008); *Poets on the Edge: an Anthology of Contemporary Hebrew Poetry* (SUNY Press, 2008); *With an Iron Pen – Twenty Years of Hebrew Protest poetry* (SUNY Press, 2009); *Plague of Thirst* (Kalligram, 2016).

Rina Jean Baroukh is a Phd student in the department of Hebrew Literature at Ben-Gurion University of the Negev. She is writing her dissertation on the prose of Shimon Adaf. Her article about *The Rose of Judea* trilogy by Shimon Adaf will be published soon in *Dapim Le-Mehkar Be-Sifrut* journal.

Ruth Lubashevsky is a PhD candidate at the Department of the Arts in Ben-Gurion University of the Negev, Israel. Her research focuses on stylistic analyses of Northern Renaissance art by means of image processing. Her article titled "'Cranach Inc.'-A Case Study Determining the Nature and Extent of Lucas Cranach the Elder's Involvement in his Industrious Workshop using Image Processing, " was published in *SPIE, Optics for Arts, Architecture, and Archaeology*. She is currently writing her dissertation on Northern Renaissance Grisailles, and preparing a paper titled ""Deadmouth:" Mouth-Covered Women in Northern Renaissance Ceremonial Acts."

Romi Mikulinsky lectures at Bezalel Academy of Arts and Design in Jerusalem. Her dissertation, written at the University of Toronto, was dedicated to photography, memory, and trauma in literature and film. She studies the present and future interfaces of texts, images, codes, and networked communities. Co-authored with Prof. Sheizaf Rafaeli, her book *Digital Clutter: Topics in Digital Culture* will be published in 2017.

Ronit Milano is a senior lecturer at the Department of the Arts in Ben-Gurion University of the Negev in Israel. Her book, *The Portrait Bust and French Cultural Politics*, was published in 2015. In the past few years, following a post-doc in Harvard University, her research has been focused on contemporary art. Milano has published articles concerning the political, economic and inter-cultural context of contemporary art and the new role of museums in a changing arena. She is head of the Museum Studies program in her department.

Susan Nashman Fraiman is a lecturer at the Rothberg International School of the Hebrew University. In the past, she served as the Collection Manager of the Yad Vashem Art Museum, as an adjunct lecturer at the Hebrew College in Newton, Massachusetts, and as a curator in the Jerusalem Biennale for Contemporary Jewish Art in 2015. Her research focuses on Jewish women and Jewish art of the modern period.

Sandra Valabregue-Perry is an independent scholar and a painter specializing in medieval Kabbalah and philosophy. She holds an M.A. degree in Philosophy from the Sorbonne in Paris, and a PhD In Jewish Thought from the Hebrew University in Jerusalem. In the past years, she has been teaching at UCLA, Yale University, Queens College, Columbia University, Ben Gurion University of Negev, Safed College, the Hebrew University and the Israel Museum in Jerusalem. She has published various articles and her 2010 book is titled *Concealed and Revealed: "Eyn Sof" (Infinity) in Theosophic Kabbalah* (Hebrew).